I0488809

TELEVISÃO
SONHO

rede planetária de
televisão cultural

para a paz mundial

emanuel dimas de melo pimenta

título: **Televisão Sonho - Rede Planetária de Televisão Cultural para a Paz Mundial**
autor: Emanuel Dimas de Melo Pimenta
ano: 2015
Televisão, projeto humanitário

www.emanuelpimenta.net

de acordo com o acordo ortográfico de 1991

ISBN-13: 978-1522723882
ISBN-10: 1522723889

capa: © Emanuel Dimas de Melo Pimenta

*Para Durval de Noronha Goyos
e João Barba*

*muitos agradecimentos a todos envolvidos
neste projeto, em especial à Vera Dantas*

page

em inglês

Você pode dizer que sou sonhador, mas não estou só. Espero que algum dia você se junte a nós. E o mundo se tornará um.

John Lennon

Menino em Cabo Verde. Foto de Emanuel Pimenta, Praia, Cabo Verde, 1999. Walden Zero Collection.

Eu estava em Londres no início dos anos 1990, almoçando com Peter Band, então um bom amigo. Ele me contou que a OPEC *Organization of the Petroleum Exporting Countries* tinha fundos para projetos humanitários.

Ele disse que tinha um amigo naquela organização, que eles estavam em busca de novas idéias, que fossem importantes para o mundo, e Peter tinha lhes falado sobre os meus trabalhos, principalmente na área de arquitetura experimental.

Ele me desafiou a pensar em algo.

Fiquei um pouco perdido no momento. Parecia-me algo demasiadamente grande. Pedi alguns dias para refletir.

Naquela época eu era um dos coordenadores do primeiro festival de vídeo arte e arte eletrônica do mundo, em Locarno, Suíça, junto com René Berger, Rinaldo Bianda, Lorenzo Bianda, Nan June Paik, Edgar Morin, Madeleine Gobeil, Bill Viola e Pierre Restany entre outros.

Eu tinha participado, na qualidade de um dos coordenadores do evento, de uma apresentação de Tim Berners-Lee sobre a World Wide Web - o lançamento da www que revolucionaria o mundo.

Minha filha Laura estava nascendo.

Foi o momento em que aconteceu a primeira guerra do Golfo.

Era claro para mim que Samuel Hundington tinha razão quando dizia de um conflito de civilizações - o quê era e continuou sendo politicamente incorreto.

O mundo se desenvolvia numa louca corrida para o suicídio ecológico, para uma brutal assimetria social, e naturalmente para aquilo que viria a se revelar com os fanáticos enlouquecidos terroristas islâmicos anos mais tarde.

Não se tratava de revoltas de adolescentes, de desequilíbrios individuais ou de pequenos grupos. Eram valores diferentes.

Não havia dúvidas - uma nova guerra mundial se intensificava, subterrânea e silenciosamente - e havia apenas uma saída: comunicação, formação humana, conhecimento, descoberta.

Lembrei-me de Marshall McLuhan quando dizia, ainda nos anos 1970, "todas as formas de violência são buscas de identidade. Quando você vive na fronteira, você não tem identidade. Você não é ninguém". Ora, o universo eletrônico colocou todos na fronteira, e especialmente os países onde a literatura não se enraizou profundamente.

Um projeto interessante seria o de estabelecer uma rede de comunicação de arte e cultura. A Internet ainda não estava suficientemente desenvolvida para isso.

A solução seria uma rede de televisão.

O objetivo inicial seriam os países africanos de língua portuguesa. Depois, gradualmente, transformar a idéia num projeto planetário.

Uma rede de televisão através da qual as pessoas conhecessem melhor a sua produção cultural, mas também a de outros países, outras culturas, outros lugares. Tudo sob o signo da liberdade, sem ideologias ou religiões.

Liberdade.

Enviei a idéia ao Peter, que a enviou ao seu amigo. A resposta chegou rápido: eu deveria preparar uma proposta concreta, pois as possibilidades de obter um apoio da OPEC eram grandes.

O apoio viria em termos de um empréstimo de longo prazo praticamente sem juros.

Isso aconteceu em 1993. Comecei a elaborar o projeto. Rapidamente, convidei um querido amigo, João Barba, grande especialista em televisão, que deu um forte apoio especialmente nos dados técnicos.

Eu e João Barba fomos a Londres discutir o projeto para o poder desenvolver com bases mais sólidas e com melhores possibilidades de obter o apoio da OPEC.

Logo compreendemos que aquele projeto era muito mais importante do que poderíamos imaginar, ele era importante a nível planetário.

Em 1996 convidei outro querido amigo para o projeto: Durval de Noronha Goyos, grande advogado, árbitro da Organização Mundial do Comércio e um dos responsáveis pelo desenho jurídico do Mercosul.

Assim, eu, João Barba e Durval de Noronha passamos a trabalhar em equipe, no sentido de aprimorar o projeto.

No dia dezessete de julho de 1996 era criada a CPLP Comunidade dos Países de Língua Portuguesa, formada por Angola, Brasil, Cabo

Verde, Guiné-Bissau, Moçambique, Portugal, São Tomé e Príncipe, e Timor Lorosae.

Decidimos apresentar o nosso projeto como a primeira rede de televisão da CPLP.

Foram mais quatro anos em desenvolvimentos e negociaçãoes até o ano 2000, quando Dulce Maria Pereira, então Secretária Executiva da CPLP, abraçou a idéia.

Mas, aconteceram os ataques terroristas de 11 de Setembro em 2001 e, tal como George W. Bush já tinha anunciado, a ordem mundial mudou, e tudo acabou.

Todo o esforço para estabelecer uma rede de cultura e arte foi subitamente cortado.

Cerca de vinte anos se passaram desde o início dos trabalhos para a implantação desse grande projeto de comunicação.

Agora, decidi publicar o projeto - como ele foi apresentado, pois eventualmente poderá ser útil a outros projetos semelhantes.

Foram excluídos os muitos gráficos, tabelas e diagramas que constituíram o projeto original.

As descrições país a país são muito semelhantes. Esta edição não pretende trazer um modelo para implantação de uma rede de televisão planetária, até porque rapidamente a tecnologia do setor conheceu um gigantesco salto, sem contar com a emergência da Internet.

Esta edição pretende apenas ser uma referència histórica.

Este pequeno livro é dedicado ao João Barba e ao Durval de Noronha Goyos, dois queridos amigos que foram sempre incansáveis entusiastas do projeto, e sempre conscientes da sua importância em termos humanitários.

É também dedicado à minha filha Laura, que nasceu pouco antes do início dessa aventura.

Os meus sinceros agradecimentos a todos os que participaram deste sonho.

Emanuel Dimas de Melo Pimenta
Locarno. 2015

CPLP

FICHA DE PROJECTO

Nome do Projecto: Rede de Televisão Cultural e Educativa dos Países de Língua Portuguesa

Número do Projecto:

Entidade Proponente: CPLP, NORONHA Advogados (São Paulo, Brasil), Duvídeo Profissionais da Imagem (Lisboa, Portugal).

Responsável pela Acção: Duvídeo Profissionais da Imagem e CPLP.

Data: Março de 2001

Local: Angola, Cabo Verde, Guiné-Bissau, Moçambique, São Tomé e Príncipe, e Timor Leste.

Duração: indefinida.

Justificação: Na Declaração dos Chefes de Estado e de Governo da CPLP sobre Educação, de 17 de Julho de 1998, foi afirmado o "compromisso de se organizar um conjunto de acções para aprofundar a cooperação comunitária na área educacional por meio de várias directrizes entre as quais ... o reforço do ensino médio, técnico profissionalizante, ... a promoção do ensino à distância nos diversos níveis e ... ampliar a disponibilidade de fornecimento de material didáctico de natureza variada e compatível com os sistemas de ensino de cada país". Excepção feita a Portugal e Brasil, o crescimento demográfico nos países acima referidos será fortemente relevante nos próximos vinte anos, segundo dados estabelecidos pelo Banco Mundial e pela Unesco. Torna-se, assim, urgente uma estratégia que atenda com eficácia às necessidades de formação quer no âmbito especificamente relacionado ao ensino sistematizado, quer à valorização das raízes culturais de cada região.

Países participantes: Angola, Brasil, Cabo Verde, Guiné-Bissau, Moçambique, Portugal, São Tomé e Príncipe, e Timor Loro Sae.

Rede de Televisão Cultural CPLP

A CPLP

A CPLP é uma organização internacional criada no dia 17 de Julho de 1996 com sede em Lisboa e consistindo dos seguintes Estados Membros: Angola, Brasil, Cabo Verde, Guiné Bissau, Moçambique, Portugal e São Tomé e Príncipe.

A CPLP tem como objetivo promover as relações político-diplomáticas entre os seus Membros, especialmente com relação à reforma das Nações Unidas e do Banco Mundial, às relações entre União Européia e Mercosul, e a realização do Encontro Europa-África.

A CPLP é também dedicada à cooperação, particularmente em termos econômicos, sociais, culturais, legais e técnico-científicos; assim como a projetos de promoção e difusão da língua Portuguesa, nomeadamente o melhoramento do Instituto Internacional da Língua Portuguesa e a criação de um Fundo Bibliográfico.

Todas as decisões são tomadas por consenso.

Projeto Executivo Básico

Panorama Geral

Das estimadas 778.500.000 pessoas na África, cerca de 65% vivem em áreas rurais. Apenas em Cabo Verde, um dos países subsaarianos mais desenvolvidos, cerca de 30% da população vive na pobreza, e cerca de 14% em absoluta pobreza. A pobreza é predominantemente um fenômeno rural: 70% dos pobres e 85% dos ultra pobres estão em áreas rurais.

A incidência de pobreza é mais elevada nas famílias analfabetas e dirigidas por mulheres. Cerca de 40% das famílias em Cabo Verde são dirigidas por mulheres e mais de um terço estão na pobreza.

Alguma informação estatística pode nos dar com grande clareza o cenário de tal dramática realidade:

A população na África era estimada em 778.500.000 pessoas no ano de 1998. Segundo os dados do Banco Mundial e da Unesco, até o ano de 2025 estima-se um crescimento demográfico que lançará a população Africana para cerca de 1.453.900,000 habitantes, o que representa um aumento de cerca de 90% em apenas 25 anos.

O consumo de energia é outro importante indicador das tendências de desenvolvimento no continente Áfricano. A tabela abaixo compara o consumo de algumas específicas fontes de energia na África, na Europa Ocidental e nos Estados Unidos entre 1986 e 1997. A informação está em quatrilhões de Btu:

Petróleo

	1986	1990	1993	1995	1996	1997
África	3.67	4.17	4.56	4.83	4.91	5.11
Europa	26.54	27.42	28.37	29.61	29.97	30.33
EUA	38.35	40.39	40.62	41.60	42.96	43.81

Energia hidroelétrica

	1986	1990	1993	1995	1996	1997
África	0.51	0.57	0.57	0.59	0.60	0.62
Europa	4.57	4.74	5.17	5.24	5.06	5.17
EUA	6.85	6.38	6.71	7.18	7.87	7.80

Total de Consumo Energético

	1986	1990	1993	1995	1996	1997
África	9.49	9.33	10.18	10.82	11.05	11.42
Europa	61.06	63.93	64.35	65.97	67.29	68.20
EUA	88.82	97.02	100.79	104.92	108.43	109.14

Em 1997, seguindo as mesmas fontes de dados, o PIB *per capita* nos países subsaarianos da África era de cerca de 510 dólares, contra uma média de 25.890 dólares na Europa Ocidental, Estados Unidos e Japão.

Também como um quadro comparativo, entre 1995 e 2000, as taxas de mortalidade infantil estimadas foram:

África   86 por 1.000 nascimentos

Ásia   56 por 1.000 nascimentos

Europa   12 por 1.000 nascimentos

América do Norte 7 por 1.000 nascimentos

Considerando alguns específicos países de língua Portuguesa, segundo fontes do Wistat e das Nações Unidas, de 1990, o percentual de analfabetismo é:

Entre 15-24 anos de idade:

> ANGOLA: Mulheres: N/A; Homens: N/A
>
> CABO-VERDE: Mulheres: 13,6 %; Homens: 10,1 %
>
> GUINÉ-BISSAU: Mulheres: 82,2 %; Homens: 40,3 %
>
> MOÇAMBIQUE: Mulheres: 74,7; Homens: 36,0
>
> SÃO TOMÉ E PRINCIPE: Mulheres: 25,9 %; Homens: 9,2 %

Entre 25 anos de idade e mais velhos:

> ANGOLA: Mulheres: N/A; Homens: N/A
>
> CABO-VERDE: Mulheres: 62,5 %; Homens: 34,9 %
>
> GUINÉ-BISSAU: Mulheres: 95,7 %; Homens: 77,8 %
>
> MOÇAMBIQUE: Mulheres: 93,8 %; Homens: 65,8 %
>
> SÃO TOMÉ E PRINCIPE: Mulheres: 36,9 %; Homens: N/A

O percentual de analfabetos em toda a África, entre os anos de 1995 e 1999 foi:

	1995	1998	1999
Homens	35%	32%	31.1%
Mulheres	52.9%	48.7%	47.4%

A comunicação é outro indicador interessante de desenvolvimento. Baseado em informações da Unesco datadas de 1997/1998, comparando alguns países africanos de língua Portuguesa, África em geral, Reino Unido e Estados Unidos, temos a seguinte imagem (por cada mil habitantes):

	1	2	3	4	5
Angola	-	0.3	-	4.7	34
Guiné	-	-	0.5	7.3	42
Moçambique	0.8	-	0.4	3.4	38
Grã Bretanha	193	122	30.8	527.8	1433
EUA	362	165	53.9	639.9	2093
África	N/A	N/A	N/A	14.3	N/A

1 = Computadores ; 2 = telefones portáteis; 3 = Fax; 4 = telefones; 5 = rádios

A África é um continente submerso em problemas dramáticos. Todos estes problemas humanitários têm na educação e na cultura a sua raiz mais direta e objetiva.

Os países africanos que falam português estão entre os países africanos com o menor nível de desenvolvimento.

A situação é tão dramática que a implementação da Rede de Televisão Cultural CPLP é mais do que urgente.

Objetivos do Projeto

O principal objetivo do projeto é assistir os países Africanos - começando pelos países de língua Portuguesa - no desenvolvimento do primeiro canal de televisão africano orientado para temas culturais e educacionais.

Esta rede será chamada Rede de Televisão Cultural CPLP.

O desenvolvimento de um projeto como este irá criar um pólo de irradiação com a sua implementação em todo o continente africano, constituindo uma rede africana de pesquisa e cultura, descobrindo e valorizando diferentes aspectos culturais e, consequentemente, reforçando fatores de identidade.

Com uma identidade cultural forte, espera-se alcançar níveis mais baixos de violência no comportamento social, bem como um aumento relevante do desempenho educacional em todos os seus setores.

Objetivos educativos sistemáticos são complementares à geral orientação deste projeto, que é cultural no seu fundamento.

Um grande impacto positivo também é de se esperar no que diz respeito a vários setores da economia, melhorando significativamente a qualidade geral dos serviços.

A missão do projeto é:

- Resgatar valores perdidos e informação cultural

- Reforçar elementos históricos

- Resgatar obras de arte antigas

- Dar um novo valor para novas obras de arte

- Comunicar informação prática e objetiva sobre saúde

- Comunicar informação prática e objetiva sobre agricultura

- Comunicar informação prática e objetiva sobre economia

A Rede Cultural de Televisão CPLP terá na sua filosofia, como apoio à sua missão, os seguintes fundamentos:

- trabalhar com as pessoas locais

- estar focado na qualidade

- intercomunicação

Três palavras podem definir a filosofia da Rede Cultural de Televisão CPLP: **Pessoas, qualidade, intercomunicação**.

Seguindo essas três palavras-chave da sua filosofia, uma única frase também define a sua missão: "Um Continente da Cultura".

Dois setores principais caracterizam a sua estrutura global:

1. Setor Cultural

1.1. Desenvolvendo programas sobre:

- música

- arquitetura (também vernacular)

- arqueologia

- literatura e poesia

- história

- turismo

- dança

- artes plásticas

- astronomia

- ciências em geral

- gastronomia

- saúde

- antropologia

- culturas locais

- outros

1.2. Os programas podem ser de todos os tipos, o mais importante é estar atento à comunicação. Todos os programas devem alcançar o maior número possível de espectadores, sempre com muita qualidade.

2. Setor educacional.

2.1. Educação sistematizada em curtos cursos práticos.

2.2. Séries educativas com informação prática e objetiva sobre:

- cuidados de saúde

- família

- agricultura

- administração

- serviços

- construção civil

Na medida em que a Rede Cultural de Televisão CPLP não deverá ser apoiada pelos respectivos governos no futuro, ela terá um tempo diário de programas comerciais gratuitos de até 12 horas. Esta é uma boa solução para pagar seus custos e apoiar o seu desenvolvimento.

Declaração de Missão

Melhorar os níveis gerais da realização cultural da população, reduzindo a violência em geral.

Melhorar a qualidade e a relevância das competências técnicas, vocacionais e profissionais.

Fortalecer a capacidade institucional para o desenvolvimento de políticas e implementação de programas em diferentes áreas.

A Rede Cultural de Televisão CPLP não lidará com:

- violência

- destruição da Natureza

- política

- religião (no sentido de estabelecer quaisquer julgamentos de valor)

Por outro lado, a Rede Cultural de Televisão CPLP vai trabalhar na melhoria da identidade nacional, regional e local, reduzindo a violência. Vai defender os valores humanos e a preservação da Natureza. Não vai lidar com partidos políticos, e a religião será sempre tomada como um valor universal, nunca dando a esta ou aquela religião em particular qualquer valor especial.

A Rede Cultural de Televisão CPLP é um projeto para a paz e seu objetivo não é o de competir com canais de televisão locais já instalados, mas sim o intercâmbio com eles, a promoção de uma futura rede mundial dedicada ao conhecimento e ao desenvolvimento.

Os primeiros passos

O primeiro objetivo da Rede Cultural de Televisão CPLP é instalar seus estúdios nos seguintes países:

- São Tomé e Príncipe

- Cabo Verde

- Moçambique

- Guiné

- Angola

- Timor

O próximo passo será o de estender a Rede Cultural de Televisão CPLP também à África do Sul.

Assim, estes sete países irão estruturar a rede básica de todo o projeto.

Descrição geral

As componentes físicas básicas do projeto são:

1. A criação de empresas, em cada um dos Estados da CPLP e também na África do Sul, orientadas para os objetivos acima referidos.

2. O Estado de cada país - que será o responsável e garantidor do pagamento do financiamento da OPEP - também terá uma participação nas empresas.

3. Todas as estratégias serão orientados para menores custos a longo prazo.

4. Todo o projeto será desenvolvido por fases.

5. O calendário de transmissão será definido caso a caso.

6. O projeto incluirá:

6.1. um plano completo para uma rede de equipamentos

6.2. um plano completo para uma rede de software

6.3. todos os *procurements*

6.4. equipamentos de transporte e instalação

6.5. todas as configurações

6.6. criação do sistema infográfico

6.7. formação técnica do pessoal

6.8. projetos de arquitetura e engenharia para a construção

6.9. construção dos edifícios, organizados em fases

O Projeto

Objetivos do Projeto

O objetivo central do projeto apresentado é possibilitar aos países da CPLP Comunidade dos Países de Língua Portuguesa desenvolver canais televisão orientados para temas culturais e educativos com as seguintes finalidades:

- difusão da língua Portuguesa

- formação de quadros técnicos na área de comunicação e tele-educação

- formação e informação para melhor participação dos povos da CPLP no processo de desenvolvimento global

- apoiar o projeto de implantação do Instituto de Língua Portuguesa da CPLP em Cabo Verde

O desenvolvimento de um tal projeto criará um processo de pólo de irradiação dos países de língua Portuguesa, descobrindo e valorizando diferentes aspetos culturais e, consequentemente, reforçando os fatores de identidade regional.

Como resultado de uma identidade fortalecida, espera-se uma redução dos níveis de violência no comportamento social assim como um aumento substancial das performances de educação em todos os seus níveis.

As áreas abrangidas pelo projecto serão:

- música

- arquitetura (também vernacular)

- arqueologia

- paleontologia

- literatura e poesia

- história

- turismo

- dança

- artes plásticas

- astronomia

- agricultura

- economia

- indústria

- medicina

- ciência em geral

- gastronomia

- saúde

- antropologia

- culturas locais

- outros

Aumentar os níveis de saúde e educação, reduzir a violência em geral, melhorar a qualidade e a relevância das habilidades técnicas, vocacionais e profissionais, assim como fortalecer a capacidade institucional de desenvolver políticas e implementar programas em diferentes áreas, são alguns dos resultados esperados de um tal projeto.

O projeto pretende, igualmente, servir de base para a produção massiva de campanhas de saúde, nomeadamente no combate à AIDS (SIDA), às doenças infecto contagiosas, endêmicas ou epidêmicas, através de um processo de educação global das populações regionais.

Descrição

As componentes gerais do projeto são:

1. A criação de uma empresa orientada para as finalidades acima referidas.

2. Cada Estado possuirá uma participação na respectiva empresa. Essa participação deverá girar em torno de 10% do capital social.

3. Todas as estratégias serão orientadas para a obtenção de baixos custos a longo prazo.

4. Outro período de emissão, de iguais 3 horas diárias, serão dedicados especificamente a programas de educação.

5. O projecto implicará:

5.1. a elaboração completa de plano de rede de equipamentos

5.2. a elaboração completa de plano de rede de *software*

5.3. todos os *procurements*

5.4. transporte e instalação dos equipamentos

5.5. configurações

5.6. elaboração de todo o sistema infográfico

5.7. formação técnica do pessoal

5.8. projetos de arquitetura e engenharia para o edifício

5.9. construção

Designação de infra-estruturas:

1. Estúdio – 800 m2 utilizando o conceito arquitetônico de desprogramabilidade.

2. Estúdio integrando edifício com um total de até 2.000 m2 de área construída.

3. Equipamento de pós-produção – pós-produção não linear, duas estações ES7.

4. ENG – 3 DVCAM Camcorders

5. Régie com 4 DXC-D30PK1 Digital Cameras – para instalação de iluminação fria.

6. Flexicart.

7. Suite de computação gráfica.

8. Sala de som.

9. MUST SYSTEM: Automatic Emmission Managing System, com informação gráfica para emissões não ativas (Este sistema tem sido utilizado por diversos canais temáticos Franceses)

Obs. Duvideo/TGT Interactiva são as representantes exclusivas deste sistema para Portugal e países Africanos.

Outros dados técnicos - a instalação é caracterizada por:

1. Rede informática com UTP level 5 cables.

2. Sistema de telefonia.

3. Mobiliário específico.

Discriminação dos dados técnicos:

Estúdio

1. Sistema de iluminação fria.

2. 4 canais de câmera compostos por:

2 DXC-D30PK Sony Digital Cameras

4 J18 Canon Zoom Lens

4 CA-537P Sony Camera Adaptors

4 DXF-50 Sony Studio Viewfinder's

4 CCU-M3P Sony Control Camera Unities

4 RM-7P Sony remote CCU contols

4 50M Sony Multicore Camera Sets

4 Canon Focus and Zoom control sets

4 Headphones de intercomunicação

4 Special tripods

1 charriot plano e curvo

1 Teleprompter com 2 sistemas de leitura (monitores para locutores individual)

1 mesa de controlo de iluminação com memória – ARRI Diferentes microfones com diferentes características direccionais

Componentes audio

Componentes vídeo

Sistema Geral de Intercomunicação

Video Régie:

1. 1 Recorder/Player BETACAM equipment

2. 1 video digital mixer with 12 tracks

3. 1 DVE equipment of digital effects for 3 channels

4. 1 TBC Frame Synchronizer

5. 2 Oscyloscopy and Vectorscopy for control of cameras

6. 2 Synchronizer Generators with Changeover

7. 1 Intercommunication system with 4 places

8. 1 Audio and Video Matrix with remote controls

9. Video distributers

Audio Régie:

1. 1 audio mixer with 24 tracks

2. 2 audio monitors

3. 1 cd player

4. 1 cassette deck

5. 1 DAT

6. 1 Mini-Disc

7. 1 Level Detector for Audio Stereo

8. Stereo Audio Distributors

9. Audio compressors

10. Audio effects

11. Microphones

Video Post-Production:

1. 3 Sony ES-7 DVCAM/AVID hybrid non linear edition stations

2. 1 Conventional BETACAM Edition Suite 2:1

3. Multitrack audio (8) on hard disc, Sound Scape type

3 ENG SETS:

1. Sony DSR-300 PK DVCAM Camcorder Digital Compact

2. Report projectors

3. Vários materiais de suporte (Baterias etc.)

Todo o material acima descrito é suportado pelos dispositivos periféricos funcionais.

EDIFÍCIO

O terreno será doado por cada respectivo Estado.

- Área total de construção: até 2.000 metros quadrados

- Área do terreno: preferenciamente de 15.000 a 20.000 metros quadrados

Prazos gerais

Projetos de arquitetura e engenharia: 6 meses*

Construção: de 6 a 10 meses*

Instalação dos equipamentos: 3 meses após a entrega do edifício**

Preparação dos primeiros operadores: 2 meses

não considerado eventuais prazos burocráticos junto às autoridades locais e dependendo das características de construção.

** *o início de operações poderá acontecer antes da entrega do edifício, de forma parcial e provisória, em outro local, de forma a acelerar o cronograma*

CUSTOS

O custo total de cada unidade (turn-key) é de USD.11.600.000 (onze milhões e seiscentos mil dólares Americanos), pagos em três anos, sendo financiados pela OPEC por um prazo de 17 (dezessete) anos com um prazo de carência de 5 (cinco) anos.

O financiamento da OPEC cobre os seguintes pontos:

- projectos de arquitetura e engenharia

- construção e acompanhamento das obras

- período de funcionamento de 5 anos

- compra de equipamentos

- instalação

- configurações

- treinamento de pessoal

- e demais ítens previstos na descrição do projeto (ver abaixo)

Assim, o início dos pagamentos do financiamento à OPEC acontecerá cinco anos após o início do recebimento do financiamento. A taxa de juros aplicada pela OPEC será extremamente baixa.

Estimativa de implantação total do projecto.........1 ano

Estimativa para o break-even point...................5 anos

Período de financiamento de funcionamento.........5 anos

O projecto (turn-key) para uma estação de televisão cultural em Cabo Verde significa:

- elaboração do plano de rede de equipamentos

- *procurement* completo

- transporte e instalação dos equipamentos

- ajustes e configurações dos equipamentos

- projetos de arquitetura e engenharia

- construção do edifício

- acompanhamento das obras

- elaboração do sistema infográfico

- treinamento do pessoal

- supervisão do funcionamento e gestão

VIABILIDADE

Para a contratação de um sistema semelhante a outra televisão, por um período de transmissão de 6 horas diárias, o custo do mercado gira em torno de USD.500.000* por mês. O que significa um total de cerca de USD.6.000.000 por ano. Durante 16 anos, já descontado o primeiro ano de implementação, o custo de contratação de outro canal para as mesmas seis horas diárias é, portanto, de cerca de USD.100.000.000 – valor muito superior ao do projeto apresentado.

Esses custos são baseados nos preços praticados atualmente em Portugal, como por exemplo o programa *Hora Viva - Segurança em Directo*, transmitido diariamente, excluindo os fins de semana, das 7 às 10 horas da manhã.

Outro exemplo, o Canal Notícias Lisboa CNL, custou apenas no primeiro ano de instalação cerca de USD.10.000.000$00, sendo que o equipamento e as instalações são alugados e não de propriedade da emissora.

* valores tomados junto ao mercado Português, que se encontra abaixo do mercado médio Europeu em cerca de 30%.

Os primeiros dez meses serão dedicados a:

- design da estrutura de equipamentos e software

- elaboração dos projetos de arquitetura e engenharia

- *procurement* e aquisição de todo o equipamento

- construção e supervisão de obras

- instalação e configuração do equipamento

- início de operações

Assim, o primeiro ano será predominantemente dedicado para a implantação do projeto. Todavia, como forma a acelerar o cronograma, poder-se-á dar início antes do período previsto a algumas das operações da nova estação de televisão, provisoriamente em outro edifício.

As estimativas apontam para que a partir de 5 anos de funcionamento regular, esse canal televisivo cultural e educativo exportará uma grande quantidade de serviços e vídeos para África, Europa e América do Sul.

A viabilização econômica do projeto também acontecerá com a comercialização de diversos produtos:

1. A venda e produção de programas de natureza cultural para diversas

instituições, tais como:

- BBC

- Discovery

etc.

Em todo o mundo, redes comerciais orientadas para programas de natureza cultural têm sido criadas. Várias estações de televisão em diversos países compram regularmente programas de produtores espalhados por todo o mundo. Não apenas, várias organizações internacionais, tais como o Banco Mundial e as Nações Unidas entre várias outras, necessitam de programas de educação pública, tais como programas de sensibilização e educação quanto a doenças de caráter endêmico ou epidêmico, como a AIDS (SIDA), a malária, a tuberculose etc. Ou, de programas que focalizem o trabalho no campo, as novas técnicas de agricultura etc.

Essas instituições mundiais vêm realizando um grande esforço no sentido de desenvolver programas de televisão com objetivos humanitários.

O preço para cada campanha institucional, com uma média de produção de 10 (dez) filmes, é de cerca de USD.15.000$ e a capacidade do Canal Cultural e Educativo CPLP, no prazo de seis meses a um ano, será de duas campanhas por mês.

2. A venda e produção de programas de natureza não-cultural para outras estações de televisão, tais como:

- telenovelas

- programas de auditório

etc.

Com uma única venda de um programa de 3 horas diárias, para cinco dias por semana, por um preço equivalente ao de Portugal – e, portanto, abaixo em cerca de 30% da média Européia, e com a vantagem de não possuir elevados custos de transporte – o projeto para a televisão cultural e educativa faturaria o dobro de todo o investimento realizado.

O preço para esse tipo de programas de televisão é de:

Programas de auditório USD.12.500$ por programa

Talk-shows USD.4.000$ por programa

Entrevistas USD.4.000$ por programa

A capacidade de produção, num prazo de seis meses a um ano, desses programas pelo Canal Cultural e Educativo será de:

Programas de auditório: 6 por mês

Talk-shows: 20 por mês

Entrevistas: 20 por mês

O preço das telenovelas é muito superior. O preço de cada telenovela completa, com cerca de 120 capítulos, é de cerca de USD.2.500.000 e a capacidade do Canal Cultural e Educativo será de uma telenovela por ano.

3. Abre-se a possibilidade da utilização comercial do tempo de transmissão para além das seis horas especificamente reservadas ao canal cultural e à educação.

O dia convencional de televisão é de 18 horas, das quais apenas 6 seriam não comerciais. Teríamos, portanto, 12 horas de espaços de natureza não cultural que poderiam ser livremente comercializados.

Mesmo o período de 6 horas – dividido entre 3 horas dedicados à cultura e 3 horas à educação – possui um grande potencial para a captação de patrocínios.

Cada hora comercial de transmissão pode ter até 12 minutos de publicidade. O preço por cada 30 segundos de publicidade é:

Horário nobre das 18:00 às 23:00 cerca de USD.400

Horário normal restante cerca de USD.200

Concentrando o horário comercial na faixa nobre, com 4 horas de transmissão nesse período, teríamos:

Televisão educativa: da 7:00 às 9:00 e das 15:00 às 16:00

Televisão cultural: das 16:00 às 19:00

Televisão comercial das 9:00 às 15:00 e das 19:00 à 1:00

Assim, o período comercial compreenderia 6 horas em horário normal, 4 horas em horário nobre e 2 horas em horário normal.

Sendo 12 minutos por hora de publicidade, o período comercial poderia

compreender:

Horário nobre: 48 minutos, 24 filmes = USD.9,600$ por dia

Horário normal: 96 minutos 48 filmes = USD.9,600$ por dia

Total...........USD.19.200$ por dia ou USD.576.000$ por mês

4. O arrendamento dos estúdios para equipes de televisão de outros países.

O período de arrendamento de um estúdio de televisão é de 12 horas.

Cada período tem um preço de cerca de USD.4.000$. Após o primeiro ano, poder-se-á arrendar o estúdio em até 12 horas por cada período de 3 dias. Assim, o arrendamento por mês, nesse período, poderá gerar receitas no valor de USD.120.000$ por mês.

5. Serviços de apoio a outras cadeias de televisão em todas as áreas, incluindo séries, novelas, mini-séries, docdramas etc.

Cada equipe fora do país tem um custo de cerca de USD.1.000 por dia (duas pessoas), para além dos custos de deslocação, estadia e alimentação. Nos primeiros meses poderão ser organizadas até 5 equipes para trabalhos no exterior. A capacidade de cada equipe é de 20 dias por mês. Assim, o apoio a outras redes de televisão, se tomado na sua potencialidade, poderá representar em até 100 dias por mês de trabalhos, significando cerca de USD.100.000$ por mês de faturamento.

6. A realização de colóquios, seminários, videoconferência e encontros de diferentes modalidades.

O edifício sede da televisão cultural e educativa poderá possuir um auditório médio, com plenas condições para receber seminários, colóquios e encontros das mais diversas naturezas. A realização de tais seminários, videoconferências etc., não é apenas importante para o aumento da faturamento, mas também atrairá para Cabo Verde especialistas das mais diferentes áreas, sendo um elemento importante para o desenvolvimento de outros projetos e para o turismo.

Os encontros, seminários e colóquios reúnem, em média, cerca de 300 pessoas por cada evento. O preço – excetuando-se alimentação, estadia e transporte – por cada participante é de cerca de USD.50 por dia. Haverá uma capacidade para até quatro eventos desse tipo por mês, o que poderá representar um faturamento de cerca de USD.60,000 por mês.

O projeto poderá, ainda, viabilizar:

- cursos de formação profissional, em diversas áreas

- apoio para um centro de alta tecnologia multimédia

- apoio para um centro cultural multimédia

Mas, é importante ressaltar que embora este novo canal cultural e educativo de televisão certamente será um grande sucesso comercial, este não é o seu principal objetivo.

Assim, a viabilidade do projeto deve ser considerada não apenas em relação ao retorno financeiro, mas principalmente à urgente necessidade de um projeto desta natureza.

PROJECT REPORT

OF

ANGOLA CULTURAL TELEVISION NET-WORK

AT

ANGOLA

AFRICA

PROMOTERS

REPUBLIC OF ANGOLA

AFRICA

ANGOLA TELEVISION NETWORK

INTRODUCTION

CHAPTER - 1

INTRODUCTION

1.1   Description of the organisation through which this project is being taken up.

The CPLP is an international organisation, created in 17 of July of 1996, with Headquarters in Lisbon and consisting of the following State Members: Angola, Brazil, Cape-Verde, Guinea-Bissau, Mozambique, Portugal, Sao Tome e Principe and East Timor (Timor Lorosae).

The CPLP has as its objective the politic-diplomatic relationships between its Members, mainly referring to the United Nations and the World Bank, the relationship EU/MERCOSUL and the accomplishment of the Europe-Africa Summit.

The CPLP is also dedicated to the co-operation, particularly in the economic, social, cultural, legal and technical-scientific matters; as well as to the projects of promotion and broadcasting of the Por-

tuguese Language, nominated to the improvement of the International Institute of the Portuguese Language and the creation of a Bibliographical Fund.

All its decisions are taken by consensus.

2.2   Description of the main reasons for setting up an African Cultural Television Network.

Africa is a continent submerged into dramatic problems. All these humanitarian problems have in the education and culture their most direct and objective root.

The African countries speaking Portuguese are among the African countries with the lowest level of development.

The whole situation is so dramatic that the implementation of the CPLP Cultural Television Network is more than urgent.

It is hoped that with this Television Network the people shall have better understanding of various diseases and other problems of life. e.g. AIDS is prevalent disease in Africa. But as the people are illiterate the information about the disease cannot be spread by any other method. In such cases the Television Network becomes a very effective audio-visual means of communication.

The project also intends to be a base of a massive production of health campaigns, specially to reinforce the combat against the AIDS, the infect-contagious diseases- endemic or epidemic-through a process of global education of the local and regional populations.

The CPLP Cultural Television Network is a project for peace and its objective is not to compete with local television channels already installed, but to interchange with them, promoting a future global web dedicated to knowledge and development.

1.03 Description of the main objects of the project.

The major objective of the project is to assist the CPLP countries to develop the first African television network channel oriented to cultural and educational themes with the follow orientation:

- diffusion of the Portuguese language
- education in all areas
- reinforcement of the local and regional cultures
- formation of technical personal on communication and tele-education

- formation and information for a better participation of the CPLP nations in the process of global development

\-   to support the implantation of the Institute of Portuguese Language

The development of such a project will create an irradiation pole from the countries with the project implementation to the whole African continent, constituting an African network of research and culture, discovering and valorising different cultural aspects and, consequently, reinforcing identity factors.

With the stronger cultural identity, it is expected to have lower levels of violence in social behaviour as well as a relevant increase of the educational performance in all its sectors.

Systematic educational objectives are complementary to the general orientation of this project, which is cultural in its fundament.
A very large positive impact is also to be expected with respect to several sectors of the economy, greatly improving the overall quality of services.

1.4   Description of the mission status of the project.

The mission of the project is:

\-   to rescue lost values and cultural information

- to reinforce historical elements
- to rescue old artworks
- to give a new value to new artworks
- to communicate practical and objective information on health
- to communicate practical and objective information on agriculture
- to communicate practical and objective information on economy

1.05 Description of the philosophy of the project.

The CPLP Cultural Television Network will have in its philosophy, as a support to its mission, the follow fundaments:

- to work with local people
- be focused on quality
- intercommunication

Three words can define the CPLP Cultural Television Network philosophy:

People, quality and intercommunication

Following these three key words of its philosophy, a single sentence also defines its mission: "A Continent of Culture".

Two main sectors characterise its global structure:
1. Cultural Sector

1.1.   To develop programmes on:
-    Music
-    Architecture (also vernacular)
-    Archaeology
-    Literature and poetry
-    History
-    Tourism
-    Dance
-    Plastic arts
-    Astronomy
-    Science in general
-    gastronomy
-    health
-    anthropology
-    local cultures
-    others

1.2   The programmes can be of all types, the most important is to be attentive to communication. All programmes must reach the highest possible number of spectators.

2. Educational sector.
2.1   Systematic education in short practical courses.
2.2   Educational series with practical and objective information on:
-    health care
-    family
-    agriculture
-    managing

- services
- civil construction

As the CPLP Cultural Television Network must not be supported by the respective governments in the future, it will have a daily time of free commercial programmes up to 12 hours. This is a good solution to pay its costs and support its development.

1.06 Description the magnitude of the project.

The project proposes a Television transmission time of eight hours per day in the first year. It will increase by two hours per day in the second and subsequent years.

The project also envisages training of local people in all the related areas of Television. It is proposed to develop a local pool of talent which can be utilised not only in this particular country but also in other countries of Africa.

The total cost of the project of the Angola Cultural Television Network is US$.26.500.000. Out of this it is proposed to obtain a soft loan of US$.22.000.000. The remaining amount of US$.4.500.000 shall be contributed by the Government of Angola in terrain for the installation of the respective building, infrastructures and logistic support.

For the purpose of the Angola Cultural Television Network a company shall be formed with the participation of the Government. The Contribution of the Government in the form of share capital shall be US$.4.500.000.

The term loan shall be on soft terms. The total period of the loan shall be 23 years. Out of which five years shall be a period of moratorium. The entire loan shall be repaid in the remaining 17 years, in 17 equal yearly instalments.

The Government of Angola will give the sovereign guarantee for the referred loan.

ANGOLA TELEVISION NETWORK

THE COUNTRY

CHAPTER - 2

THE COUNTRY

2.01 Description the level of development of Angola.

Angola is an economy in disarray because of more than 20 years of nearly continuous warfare. Despite its abundant natural resources, output per capita is among the world's lowest. Subsistence agriculture provides the main livelihood for 85% of the population. Oil production and the supporting activities are vital to the economy, contributing about 50% to GDP.

Notwithstanding the signing of a peace accord in November 1994, sporadic violence continues, millions of land mines remain, and many farmers are reluctant to return to their fields. As a result, much of the country's food must still be imported. To take advantage of its rich resources-gold, diamonds, extensive forests. Atlantic fisheries, arable land, and oil deposits-Angola will need to implement the peace agreement and reform government policies. Despite the high inflation and political difficulties, total output grew an estimated 9% in 1996, largely due to increased oil production and higher oil prices.

OFFICIAL NAME: Republic of Angola
CAPITAL: Luanda
SYSTEM OF GOVERNMENT: Multiparty Republic
AREA: 1,246 700 Sq Km (481,354 Sq Mi)
ESTIMATED 2000 POPULATION: 12,897,000

LOCATION & GEOGRAPHY: Angola is located on the western coast of South Africa. It is bound by Namibia to the south Zambia to the east, Democratic Republic of the Congo (Zaire) to the north and north-east, and the Atlantic Ocean to the west. A separate province of Cabinda is enclosed by the Congo. A sparsely watered coastal plain extends along the coast and rises towards the interior which contains irregular terraces that form sub- plateaux. The central plateau accounts for around 66% of the land area and has numerous rivers which run into basins of the Congo and Zambezi Rivers that in turn flow to the Atlantic Ocean. The north-western region of the central plateau and the enclave of Cabinda are covered by equatorial jungles while the southern regions and coastal plain are semi-arid. The Namib Desert occupies the coastal plain above Mocamedes. Major Cities (pop. Est); Luanda 1.000.000, Huambo 203,000, Beneguela 155,000, Lobito 150.000 (1983). Land use; forested 42% pastures 23%, agricultural-cultivated 3%, other 32% (1993).

CLIMATE: Angola has a tropical climate with regional variations from a moderate tropical climate to a desert climate depending on the location. The prevailing winds are predominantly from the west, south-west and south. Average temperature ranges in Luanda are from 18 to 23 degrees Celsius (64 to 73 degrees Fahrenheit) in August to 24 to 30 degrees Celsius (75 to 86 degrees Fahrenheit) in March.

PEOPLE: Almost, the entire population is Bantu origin with various and numerous tribal groups. The principal ethnic majority are the Ovimbundu who alone represent 37% of the population while the Mbundu, Lunda, Chokwe, Nganguela, Ovambo, Herero, Kangala, Humbe, Luvale, Bunda, Luchazi, Kwandare and Cuanhama account for 38% of the population. The Ovambo and Herero are nomadic cattle herders that regularly migrate across the Angola-Namibia border.

DEMOGRAPHIC/VITAL STATISTICS: Density; 8 persons per sq km (21 persons per sq mi) (1991). Urban-Rural; 26.8% urban, 73.2% rural (1988). Sex Distribution; 51.1% male, 48.9% female (1990). Life Expectancy at Birth; 42.9 years male, 46.1 years female (1990). Age Breakdown; 42% under 15, 28% 15 to 29, 17% 30 to 44, 10% 45 to 59, 3% 60 and over (1990). Birth Rate; 47.2 per 1,000 (1990). Death Rate; 20.2 per 1,000 (1990). Increase Rate; 27.0 per 1,000 (1990). Infant Mortality Rate; 137.0 per 1,000 live births (1990).

RELIGIONS: Mostly Christians with around 70% of the population Roman Catholic while 20% are Protestant. Although, a large number still adhere to traditional tribal beliefs.

LANGUAGES: The official language is Portuguese, although the national language is Bantu with the number of dialects spoken as varied as the Bantu sub-tribes.

EDUCATION: Aged 25 or over and having attained: N/A. Literacy; literate population aged 15 or over 1,196,000 or 28% (1980).

CURRENCY: The official currency is the New Kwanza (NKz) divided into 100 Iwei.

ECONOMY: Gross National Product: USD $6,010,000,000 (1989). Public Debt; USD $7,727,000,000 (1993). Imports; USD $1,347,000,000(1997).Exports;USD$3,427,000,000 (1991). Tourism Receipts; USD $20,000,000 (1993). Balance of Trade; USD $1,565,000,000 (1994). Economically Active Population; 4,166,000 or 40.3% of total population (1991). Unemployed; N/A.

MAIN TRADING PARTNERS: Its main trading partners are Portugal, Cuba, Germany, the US, the UK, Canada, Japan and the former USSR.

MAIN PRIMARY PRODUCTS: Asphalt, Bananas, Cassava, Citrus Fruits, Coffee, Copper, Cotton, Diamonds, Fish, Gypsum, Iron Ore, Limestone, Maize, Manganese, Oil, Palm Oil, Phosphates, Salt, Sisal,

Sugar Beets, Sweet Potatoes, Timber.

MAJOR INDUSTRIES: Agriculture, Cement, Chemicals, Fishing, Food Processing, Sisal, Timber, Tobacco.

MAIN EXPORTS: Coffee, Crude Oil, Diamonds, Fish, Maize, Oil, Petroleum Products, Sisal, Timber, Tobacco.

TRANSPORT: Railroads; route length 2.789 km (1,733 mi) (1998), passenger-km 326.000.000 (203.000.000 passenger-mi) (1998), cargo ton-km 1.720.000.000 (1.178.000.000 short ton-mi) (1988). Roads; length 73,830 km, (45.876 mi) (1986). Vehicles; cars 122.403 (1989), trucks and buses 44,000 (1989). Merchant Marine; vessels 111 (1990), dead weight tonnage 122.403 (1990). Air Transport; passenger- km 975.000.000 (606.000.000 passenger-mi) (1985), cargo ton-km 33.900.000 (23.218.000 short ton-mi) (1985).

COMMUNICATIONS: Daily Newspapers; total of 4 with a total circulation of 500.000 (1994). Radio; receivers 450.000 (1994). Television; receivers 50.000 (1994) Telephones; units 53.300 (1993).

MILITARY: 82.000 (1995) total active duty personnel with 91.5% army, 1.8% navy and 6.7% air force while military expenditure accounts for 23.9% (1986) of the Gross National Product (GNP).

ANGOLA TELEVISION NETWORK

THE PROJECT

CHAPTER - 3

THE PROJECT

3.1   Description of the initial steps to be taken for the implementation of the project.

The first objective of the CPLP Cultural Television Network is to install its structures in the following countries:

- Sao Tome e Principe
- Cape Verde
- Mozambique
- Guinea-Bissau
- Angola

\-   Timor

Thus, these six countries will structure the basic network of the whole project.

Angola, Mozambique and Cape Verde will form a FIRST GROUP; Sao Tome e Principe, Guinea-Bissau and East Timor will form a SECOND GROUP; and, finally, Brazil and Portugal will form a called REFERENCE GROUP.

However, the project will start at the same time in all countries. A more precise description of the meaning of each group is made below.

The basic physical components of the project are:

1.   The creation of companies, in each of the States of the CPLP, oriented to the objectives above referred.
2.   The State of each country – who will be the responsible and guarantee of the payment of the financing – will also have a participation in the companies.
3.   All strategies will be oriented to lowest costs in long term.
4.   The whole project will be developed by phases.
5.   The timetable of transmission will be defined case by case.
6.   The project will include:

6.1. a complete plan for an equipment network.

6.2. a complete plan for a software network.

6.3. all procurements

6.4. equipment transportation and installation

6.5. all configurations

6.6. creation of the infographic system

6.7. technical formation of the personal

6.8. architecture and engineer projects for the building

6.9. construction of the building organised in phases

1.2 Description of the procedure by which the necessary programmes will be produced for transmission on the channel.

The television programs are known as software in this field. The unit is planning transmission of three hours of educational programs, three hours of cultural programs and two hours of commercial programs per day in the first year. Thus, there shall be a total transmission of eight hours of programs per day in the first year. The total transmission hours shall increase by two hours per day in the second and subsequent years. Thus, in the second year there shall be total transmission of 10 hours per day and in the third year there shall be a total transmission of 12 hours per day. The cost of software production is assumed at US$.2.500 per hour. The amount of US$.2.500 per hour is taken based on cost of production of such programmes in Portugal.

To contract a similar system from other television channel, only for one country, for a transmission period of 6 hours daily, the cost in, market (eg. Portugal*) is of about US$.500.000 per month. This cost represents a total of about US$.6.000.000 per year. In 16 years, excluding the first year of implementation, the costs for contract the six daily hours of transmission from other television channel, in present market price, is of about US$.100.000.000. The six countries would have an estimated cost of about US$.600.000.000 in a 16 years period of time.

These costs are based on the prices presently used in Portugal, as for example the programme Hora Viva – Seguranca em Directo, transmitted every day, excluding weekends, from 7 to 10 AM.

Another example is the channel Canal Noticias Lisboa CNL. Only in its first year of installation, it was expended about US$.10.000.000, and the equipment as well as the building were not property of the channel, but rented.

In fact, however the project will be implemented by phases, after 5 to 10 years of regular work, it is predicted to have reached a similar value in incomes as the examples above.

*value took in the Portuguese market, below the average of the European prices in about 30%.

The first ten months of works will be oriented to:
-        design of the equipment and software structures
-        start the first productions
-        elaboration of the projects of architecture and engineering
-        procurement and acquisition of all equipment
-        construction and supervision of the works
-        installation and configuration of the equipment

So, the first year must be used to the implantation of the project. However, as to accelerate the chronogram, some operations should be started before this timing, temporarily installed in a different building.

The first phase will be oriented to:
-        video productions (external and internal)
-        edition
-        elaboration of cost programmes

3.3   Description of the requirements of Power, Fuel and Water. POWER
It is assumed that the unit shall use one thousand

units of power of everyday. Thus, the unit shall use 365.000 units of power in a year. The costs of power is assumed at five cents for one unit. The consumption of power shall increase by 10% every year.

FUEL

The unit shall use D.G. set for generation of electricity, when there is an interruption in the supply. The hours of disruption in the normal supply can not be predicted accurately. Hence, the hours for which the D.G. sets shall be used also can not be predicted accurately. However, it is assumed that the unit shall utilise one thousand litters of diesel per month. The cost is assumed that 33 cents for one litter. Hence, the total cost for one thousand litters shall be US$.330. The consumption of diesel shall increase by 10% every year.

WATER

The unit does not need water for fixed commercial operation. However, it shall have a strength of 50 people as employees. Water shall be required for drinking and sanitation purposes. It is assumed that 3000 litters of water shall be required everyday. The cost of obtaining this water is assumed at US$.10 everyday. The consumption of water shall increase by 10% every year.

3.4   Description of the manpower requirement of the project and description of the steps taken by the company to train the manpower.

The total staff strength shall be 17 people in the first year.

The system of modern TV Station is designed to achieve sustained operating efficiency and transmission. This, of course, entails a certain degree of sophistication in the production and transmission. However, considering the socio-economic situation in Angola, a reasonable balance has to be struck to obtain optimum performance and at the same time create gainful employment. While working out the manpower requirement for this project to be kept on direct rolls of the company totalling 50 staff and operators, the above consideration have been kept in mind.

Manpower requirement

The direct manpower required for the proposed Unit is about US$.1.800.000 per year. The manpower requirement as indicated in this chapter has been planned keeping in view the following guidelines:

Effective co-ordination among the various departments. Judicious distribution of responsibilities.

Capacity utilisation of the TV Station with optimum manpower.

Details of manpower requirement for the TV Station is given in the financial section.

In line with the prevailing practice, the Security guards, office peons and unskilled labour etc., are normally employed on contractual/casual basis, and their cost has to be included in the Factory overheads. However in this report provision has been made to employ the above staff and their salaried have been included.

The various departments proposed shall be under the direct responsibility of the Station Director. He shall be assisted by Deputy Directors, who will look after the complete TV Station and its various day to day activities and Marketing. They shall be responsible for achieving the envisaged targets and sales forecasts. They shall be assisted by a team of Managers from production, marketing and accounts.

To run this project a labour force which shall usually be composed of unskilled and skilled workers shall be employed. The first are those who do not undergo any king of specific training or education, while the latter have to do so in order to master their jobs.

When evaluating an investment project from an employment point of view, its impact on both unskilled and skilled labour has been taken into account. Not only direct employment, but also indirect employment has been considered. Direct employment refers to the new employment opportunities created within the project; indirect employment concerns job opportunities created in other projects linked with the project which is being formulated.

The implementation of large and sophisticated projects generally contributes to the development of local skills and capabilities in a country. Furthermore, they help to change traditional values, attitudes and the behaviour of the society, to build up an enterprising spirit among the people, to develop a desire for changing and improving the existing conditions of life, to introduce better work discipline and thus to change the very pattern and basis of economic development. The TV industry is already well established in Africa. Location of TV industry activities in Africa, has certain favourable factors and advantages, in setting up this type of industry, as availability of acceptable level of education among the supervisory staff adequate technical and managerial skills developed over a long period, and availability of cheap labour are assured.

The organisation structure will vary from TV Station to TV Station in the industry and as such the pattern proposed herein can be considered only as suggestive and provisional.

A well-knit organisation structure headed by a Station Director and Manager, with the supporting staff will be developed progressively during project implementation. Soon after the plant becomes operative, a good number of project staff will be absorbed in the organisation. Certain additional staff also get added to ensure smooth and efficient management of the operating unit.

The Manager Production will have a degree in communication and will be accountable and answerable to the Station Director for all TV operations including planning, production, material management, TV utilities, quality control, production cost and budgetary control, TV safety, discipline and layout relations. His main function should be to ensure achievement for quality and quantity targets of production at reasonable cost and should constantly strive to improve TV performance. In the discharge of his multifarious duties and responsibilities he will be assisted by supervisors and adequate staff for day to day activities.

His duties are to ensure that targets are maintained through effective utilisation of 3 M's viz: Machinery, Men and Materials. These targets should be translated in terms of targets for individuals under him such as supervisors, skilled workers etc. He has to ensure that the equipment at his disposal gets prompt attention on breakdown is adhered to strictly. He has to further ensure that there is a strict quality control exercised over raw materials also. He will be assisted by Deputy Manager etc. who and will report to the General Manager.

GENERAL MANAGER (Comm.) He will report to the Station Director. His main duties will be prepare sales forecasts and budgets, study and monitor the export market for the company's products and advice on ways and means of increasing sales. He will also ensure that consumer complains are solved in an appropriate manner. He will also ensure that the right materials are available for production at the right time.

He will be responsible for all aspects connected with the export procedures and keep the management updated on all matters relating to the Govt. policies on polymers and Exports and above all the world market.

COMPANY SECRETARY & FINANCIAL CONTROLLER. He will report to the Station Director and will be responsible for the departments of administration, financial planning, budgetary control, cost accounting, tax management payroll accounting, etc. and all personnel matters. He will be assisted in their duties by their respective assistants to assist in day to day activities.

Manpower planning and production

The central issue here may be one of scale. Production is staffed bye personnel, but the process may be labour-intensive or capital-intensive. Either way, planning will include:

1.   Analysis of labour supply and demand factors in relation to skills and training needs.
2.   Procedure for manpower recruitment, together with selection processes.
3.   The formation of industrial relations policies necessary for effective work place bargaining, disciplinary measures and dismissal procedures.
4.   Analysis of the effective use of human resources.
5.   Conditions necessary to maintain adequate levels of motivation. The scale of the problem is likely to be directly proportional to the method of production.

The availability of main persons is not going to be easy since TV industry is currently in its infancy in Africa.

Training needs

The selection and training of the required manpower for the proposed project has to be planned in advance.

The key personnel should be selected and trained suitably. The training would be carried out in the following manner:

Basic training on the concept of TV industry before construction begins with visits to similar TV Stations in other countries. On site training during the construction phase of the project. On job training during the commissioning phase of the project. On job training during operation of the TV immediately after commissioning. The training of the key personnel such as Station Director should be carried out in all the phases. The training of other operating personnel should be suitably carried out during the construction, erection and operation phases in addition to training them by visits to similar plants operating in other countries.

Besides training the key operating staff described above, in TV training should also be given to other employees at skilled operating level to enable them to understand the process equipment in the project and prepare them to operate a maintain their respective sections safely, efficiently and skilfully. The above training should be carried out during construction, commissioning and operating phases of the project.

Training is necessary in order to enable personnel to

acquire the skills and knowledge necessary to perform a task to an acceptable standard. The length of the training period and training methods will, of course, vary from job to job. Training is essentially a learning process, and in order that progress can be successfully monitored certain conditions are necessary.

1.   The training needs of both the individual and the organisation shall be identified and analysed.
2.   Targets and standards shall be set for the trainee, which are within his capabilities.
3.   The pace of the training programme should reflect the trainee's ability to maintain progress in properly absorbing the same.

4.   The trainee shall receive regular feedback of results. Any problem areas shall be highlighted, discussed and resolved.
5.   As the trainee progresses the amount of information provided shall be gradually reduced, thus inducing a feeling of independence and competence.

It is common place to find a wide variety of tasks in an organisation and each will require varying degrees of skill, effort and responsibility. This being so, it is inevitable that rates of pay will also vary and the differentials between the jobs will reflect their relative values. However, other factors such as local market conditions, bargaining strengths and traditions also influence a company's payment structure and a great deal of planning is required if rationalisation is to be achieved. One technique which has been successfully adopted by many companies to establish an equitable wage structure is job evaluation. In a job evaluation exercise a comparison is made of common criteria over a range of jobs, and the resulting analysis may be linked to a points allocation or job ranking system, and hence to a wage scale.

In conducting a job evaluation exercise it is important to cover a reasonable variety of tasks within the whole spectrum. For each, a job description is prepared setting out details of the duties and responsibilities undertaken by the employee together with a statement about his working conditions. Very often this task is undertaken by work study personnel since, they are responsible for determining methods of operation and evaluating the work content of the job. Each job is assessed factor by factor, resulting in a comprehensive comparative analysis.

The individual is the most important resources of any company and only people who are well trained, well motivated and adequately rewarded will provide a positive and synergistic contribution towards the company's objective and its organisational health.

In most cases, the factors, which may be weighted according to relative value, are as follows:

1.	Skill-education, experience and training.
2.	Effort-both physical and mental.
3.	Responsibility-for equipment, materials, initiative etc.
4.	Working conditions-general conditions, risk of accident and injury.

Pay policies affect not only individual employees but the whole organisation, and the rewards and objectives vary at different levels within the enterprise.

Industrial relations

An effective industrial relations policy is important, since is the system through which employees take part in decision-making and in many instances it affects the while atmosphere of employer/employee relationships. An industrial relations policy is essentially a set of rules whose determine procedures for negotiation on such matters as:

1.	Wage and salary scales.
2.	Terms and conditions of work.
3.	Disputes and grievances.
4.	Recruitment and dismissal.

5.   Other issues of mutual interest, e.g. closed shop, redundancies and joint consultation.

In order to promote an atmosphere of co-operation, and to minimise conflict, the needs of management and work people must be recognised by both sides. Trade unions exit to protect the interests of their members and improve their working conditions. Management, while aware of the pressures and constraints imposed by the trade unions, have a duty to maximise the use of resources at their disposal, which may be expressed in relation to profitability, return on investment, level of service, sales volume, market share and cost-effectiveness. The strategies adopted in attempting to solve industrial relations problems will vary from company to company, and indeed from union to union, but there is no doubt that they will be influenced by both internal and external factors.

Internal factors

1.   The attitudes of employees to management, and management to employees.
2.   The leadership style of management.
3.   The bargaining strength of both parties.
4.   The number of negotiating bodies.
5.   The prosperity of the company.

External factors

1.   The extent to which parent boards influence company management, and district officials influence or control local shop stewards.
2.   Whether or not bargaining is conducted at plant, local or national level.
3.   Government policy towards industrial relations.
4.   The economic situations nationally, locally or within the company itself.

3.5   Describe the project implementation schedule.

Implementation of this project is a challenging task and calls for meticulous planning, scheduling and monitoring to realise the project goals within the budgeted cost and time frame. The goal can be achieved by adopting modern project management techniques.

To implement this project adequately, a team of engineers and project personnel having requisite education and experience are being appointed, to whom a detailed Work Breakdown Structure (WBS) in a logical order of activities, shall be supplied shortly, keeping in view cost estimation, scheduling, and to help monitor and control of the project. It is proposed to be formulated in conjunction with the objectives of each activity and goal settings. Project shall be programmed and controlled by network analysis techniques. Before the application of the network analysis techniques, the project personnel shall be acquainted with their capabilities in saving time, resources and costs. The training of project personnel and engineers at levels shall be provided for proper control of project progress and taking of timely corrective actions to re-align these efforts to meet ore stated objectives. It is proposed to gear programming and control system, i.e. project implementation system, which will ensure an integrated approach to project implementation. Project management activities shall be determined in advance and all activities carried out be project personnel as well as those to be contracted shall be identified.

Responsibility for project implementation shall be clearly defined. The forms of project organisation range from project oriented to functional organisation, while most of the cases are combinations of the two, with certain adaptations to prevailing conditions. It is impossible to over emphasise the importance of establishing a team of a task force for implementing the project with a designated leader to co-ordinate and guide its functions.

Project manager: Shall be responsible to the Board of Directors. The project manager shall be responsible for guiding and co-ordinating the efforts of all parties engaged in implementing the project, obtaining necessary government approvals on contracts. He is to control the project organisation with the promoters as well as with other agencies and organisations interested in the project. The manager shall have some staff to assist him, especially in checking expenditures to date and determining the present and future cost overrun or under run so that the project manager can take or propose to the Board pertinent corrective measures.

The network shall cover the pre-construction phase of the project indicating major administrative processes, since experience shows that some of them have frequently involved lengthy delays. In other words, it shall include the aggregate activities to be carried out be the principal parties participating in the implementation process.

The project budget shall then be prepared. Part or periodic payments to contractors which might be made at the end of certain intervals (e.g. weekly or monthly) throughout the time horizon of project implementation, shall be made by summing up activity costs per unit of time, which may be a week & month, and computing the cumulative cost at the end of each time interval. For the activities that are in process and are contacted or sub-contracted, the assumption of a linear time cost activity relationship shall be used for the sake of simplicity. In other words, expenditures are uniformly distributed throughout the duration of the activity.

PROJECT SCHEDULE

After an investment decision is taken, the main machinery and long delivery items must be ordered out at the earliest, forming the first major step in implementation of the project. It is foreseen that an engineering consultant will be appointed for carrying out the detailed engineering including basic engineering and procurement assistance to the client. It is also assumed that reputed and experienced contractors with adequate resources viz., men, materials, tools and tackles etc. will be engaged for execution of the construction and erection work. The purchase packages for auxiliaries shall be kept minimum so as to reduce the co-ordination efforts to the minimum. A great deal of co-ordination is required for constructing/erecting the new units. This task is feasible, provided the major activities of the project are co-ordinated and completed in the duration specified to achieve the respective milestones in time.

STRATEGY FOR TIMELY EXECUTION

It is important to deploy a team of experienced personnel for project execution and select the external agencies with due care for rendering the services and supply of equipment for the project. The project activities must be identified, planned and scheduled, and the progress monitored for timely project implementation. All the inputs to the project including financial resources must be identified and their inflow planned and arranged in time.

The project must be managed professionally with necessary co-ordination among the various agencies and requisite decisions taken promptly.

Establishment of an effective monitoring procedure for progress review and co-ordination.

In short, the following key factors would constitute the broad strategy for timely execution of all activities in a pre-determined manner as per schedule shown in the bar chart as to reach at a basis of regular production.

Early selection of an effective in-house technical team (TASK FORCE) by Government of Angola, headed by a Project Manager for planning and executing the project.

i) Proper choice of external agencies such as consultants for Project Engineering., Machinery suppliers, Construction Agencies etc. keeping in view their reputation/past performance and working experience in their fields.

ii)   Adequate use of computer-based PERT/ CPM techniques for project planning, scheduling and monitoring.

3.6   Description of the initial project Flowchart.

The project implementation phase embraces the period from the decision to start the project to the beginning of the commercial production. It includes a number of stages including negotiations and contracting, project design, construction and start-up.

3.7   Description of the requirement of Land.

For this project the Government of Angola will participate with about 25.000 square meters of land. In selecting the land the following criteria should be followed.

1.   The land should be near to the main city of Luanda.
2.   The land should be properly connected by road.
3.   It should also have proximity to many end user clients.
4.   It should also meet various Government

policy of
   achieving the social objectives.

2.08 Description of the requirement of Building.

The total requirement of building is as follows.
a. Adm. Building   1000 Square Meters
b. Studios –3 (Three)   2500 Square Meters
c. Other Misc. Building   500 Square Meters

Total Square Meters: approximately 4000 Square Meters.
Cost for construction of about US$.750 per square meter (including air conditioning systems, primary electrical cabin etc.).

2.9   Description of some requirements for Plant & Machinery and their basis of selection.

The plant and machinery have been selected giving due consideration to the sophisticated nature of technology required. Detailed discussions were carried out with the foreign suppliers to ensure that the required capacities are practical with minimum capital and operating costs before the machinery were finally selected.

The machinery shall be selected from the most reputed foreign manufacturers of complete range of equipment. Keeping in view the following main factors:

a)	Past performance
b)	Existing machinery in Africa & abroad.
c)	Sales and service facilities in Africa

The main equipment is as follows.

STUDIO

1.	Cold lighting system
2.	8 Camera channels composed by :
4 DXC-D30PK Sony Digital Cameras
6 J18 Canon Zoom Lens
8	CA-537P Sony Camera Adapters
8	DXF-50 Sony Studio Viewfinder's
8	CCU-M3P Sony Control Camera units
8	RM-7P Sony remote CCU Controls
8	50M Sony Multicore Camera Sets
8	Canon Focus and Zoom control sets
8	Intercommunication Headphones

8 Special tripods

3 charriot plane and curve

2 Teleprompter with 2 reading systems (monitors for individual speakers)

2 lighting control table with memory-ARRI

Different microphones with diverse directional characteristics Audio Components

Video Components

General Intercommunication system

Video Regie

1. 3 Recorder/Player BETACAM equipment

2. 1 video digital mixer with 12 tracks

3. 1 DVE equipment of digital effects for 3 channels

4. 1 TBC Frame Synchroniser

5. 2 Oscilloscope & Vectorscopy for control of cameras

6. 2 Synchroniser Generators with Changeover

7. 2 Intercommunication system with 4 places

8. 2 Audio & Video Matrix with remote controls

9. Video distributors

Audio Regie

1. 1 audio mixer with 24 tracks

2. 2 audio monitors

3. 2 cd player

4.   2 cassette deck
5.   2 DAT
6.   2 Mini-Disc.
7.   1 Level Detector for Audio Stereo
8.   Stereo Audio Distributors
9.   Audio compressors
10.Audio effects 11.Microphones

Video Post-Production
1.   6 Sony Es-7 on DVCAM/Avid hybrid non linear edition stations

2.   2 Conventional BETCAM Edition Suite 2:1
3.   Multi-Track audio (8) on hard disc, sound scape type

3.   ENG SETS

1.   Sony DSR-300 PK DVCAM Camcorder Digital Compact
2.   Report projectors
3.   Various support materials(Batteries etc)

3.10   Description of the Water Requirements for the project.

The requirements of water, separately for various matters are given in the following table.

Circulating   : Nil
Make-up   : Nil

Process   : Nil
Drinking   : 3,000 LPD

3.11 Description of the Steam requirement for the project.

A. Steam requirements and steam balance   : N.A.

B. Capacity and type of boiler with detailed specifications   : N.A.

C. Steam and energy diagram : N.A.

D.   Total energy generated / purchased (converted into M. K. : N.A. Cal) theoretical requirement of energy (in M. K. Cal)

at the various consumption stations and expected actual requirement at these stations.

E. If alternate processed are available, comparative energy : N.A.
consumption figures for the various processes. If the
project is energy intensive, possibility of choosing alternate process in order to make the project less energy
intensive.

F. Steps proposed to be taken by the company to improve   : N.A.

energy losses efficiency and reduce energy losses (such

as power factor improvement, power load management,

optimising, illumination waste heat utilisation, etc.)

G. Scope for usage of solar / other renewable sources of energy: N.A.

H. Any other measures contemplated in the direction of : N.A.
energy conservation and management.

3.123 Some information on Compressed air, fuel, etc.

Compressed Air

(a) Requirement   : N.A.
(b) Sources   : N.A.
(c) Arrangements proposed   : N.A.
(d) Cost at site with detailed calculations : N.A.

3.13   Details of the nature of atmospheric, soil and water pollution likely to be created by the project and the measures proposed for control of pollution. Indicate whether necessary permissions for the disposal of effluent have been obtained.

There shall not be any atmospheric pollution likely to be created by the project as there is no machine which has combustion resulting into air pollution or any chemical process which may release any gases which may result into an air pollution.

However, the use of DG set shall cause a small amount of air pollution. Considering the size of DG set, the pollution is within the permissible limits.

ANGOLA TELEVISION NETWORK

THE INDUSTRY

CHAPTER - 4

THE INDUSTRY

4.01 Description of the Television Industry in general.

The television industry can be described in the following broad categories. 1. Introduction
2.   Cable Networks

3.   DTH

4.   TV-media

5.   Earnings drivers

6.   Outlook

The detailed discussion in each of the categories is given below.

Introduction

The growing popularity of TV as a communication medium has resulted in the TV media sector undergoing a rapid transformation. From the black and white days of state controlled TV Station, to the highly colourful tunes of Channel V and MTV, the medium has certainly undergone a phenomenal change. Given its popularity, percentage ad spend has also increased proportionately on this medium.

Media pie (%)

	1995	1997
TV	62.5	68.8
Radio	20.9	15
Press	16.6	16.2

Source NRS

Entry of new channels post 1991

All over the world the telecasting has witnessed entry of new channels to cater to the various needs of
world audiences. Channels have been launched in
English as well as other regional languages. In many
countries of the world till 1991, the state owned TV
Station ruled the roost, as other players were not
allowed to uplink and broadcast. However channels

such as CNN, Star TV and BBC, which were offshore companies, could circumvent these regulations and telecast their programs into any country of the world. Cable operators then relayed the same and made it available to the common man through the cable television network.

Like many other countries, the State machinery controlled television. It was used as a propaganda tool for the party in power, with the opposition always at the receiving end. The customer had very little choice. The first steps towards more user choice began during the 1980s, which had to be telecast to a wider audience. TV Station used satellite channels for the telecast and the T.V. network was launched as an international channel.

The sports telecast by Channel 9 in 1985 and the Gulf War in the late eighties all played small but important cameos in educating the international viewer. With liberalisation in 1992 and crumbling tariff barriers televisions (read as colour TVs) became more easily available. The media revolution had started.

Major satellite channels avidly watched by viewers are Star TV, Sony TV, Home TV, BBC & CNN. There are other regional language channels which are major players in their respective territories.

Most of the channels that could not attain popularity rapidly suffered, since their advertisement earnings were not sustainable. The first round of the media wars is over. Management changes, i.e. original promoters selling out to new management with deeper pockets, has become the order of the day. Alliances like the famous ESPN Star Sports arrangement also made headlines. Given the global trends of mergers and acquisitions, further consolidation is likely. Alliances and mergers make sense when the partners complement each other, like BBC and Discovery launched Animal Planet, CNBC and ABNI came together to launch a business channel called CNBC Asia.

Cable Networks

Antennas set up by either the end user or the cable operator receives the signals transmitted by the satellite. Local cable operators lay their own cables, set up control rooms, which can telecast 40 or more channels over a limited area. They charge the household a one- time connection charge of about US$.10 per point and a recurring monthly charge ranging between US$.1 to US$.5.

Initially, this was done in a very unorganised manner. The business required local knowledge and contacts, so every locality had its own cable operator. Collection was critical for the cable operator. For the end user, quality of telecast and a complete lack of standards became an issue. This lead to a shakeout and the formation of cable companies with money power which in turn tied-up with the local and small cable operators. Cable companies charge about US$1 per month to the local cable operators and support them with training and other infrastructure inputs. The business is immensely capital intensive and takes a long time to break even.

In many countries the operations of cable operators are regulated under the Cable TV Act which ensures that pornographic materials and other materials which are against culture and values or are detrimental to national interests do not get telecast. Recently this act has been amended to include foreign channels also.

Direct To Home

DTH is a new technology that circumvents the cable operators by directly delivering a bundle of channels to the end user. DTH involves transmission of encoded audio/ video signals (Ku band) via satellite. The end user needs an antenna to receive the signals and a decoder (set top box) to unscramble the encrypted signals. DTH services elsewhere in the world are Echostar and DirecTV (USA) and BskyB (Europe). Rupert Murdoch of Star TV fame owns BskyB.

The size of the antenna in DTH will be 1.5-2 ft in diameter, making it easy to install and transport. In conventional cable, since signals are in C band, an 8ft- diameter antenna is needed. The basic difference in the business model is the hardware costs in DTH. In a cable system, the user pays a one time connect fee and monthly rentals, while in DTH he has to invest in hardware.

The antenna will cost about US$200-300 and decoder will cost about US$200. The African viewer might be reluctant to incur such heavy installation costs. Quality of telecast in DTH is superior to Cable TV and viewer can receive up to 200 channels.

DTH will result in restructuring of the cable television industry. It will become imperative to have cash reserves to withstand the technology threat. Up gradation to fibre optic backbone will become necessary. A fibre optic network will cost about US$0.5mn per km as compared to US$0.1mn per km for coaxial cable. The stage is now ripe for consolidation.

TV Media features

In the Broadcasting business, it is only the industry leader who makes sizeable profits. The business is a game of asymmetrical payoffs. For instance, the top 5 channels account for 90% of ad spend.

Urbanisation and TV penetration is related. This may be due to the popularity of cable television that has resulted in increased colour TV sales. Rural penetration is low, although growing at a fast pace, because of dearth of specific program content to cater to that segment.

Liberalisation has resulted in the world viewer becoming more aware and conscious. This has resulted in the customer having more choice with the entry of a number of companies in different segments. Competition has resulted in companies increasing their marketing spend significantly.

Popularity of TV media is becoming higher. Increasing TV penetration leads to a reallocation of advertisement budgets with higher allocation for television at the cost of other medium.

TV channel operators use different business models to generate revenues. The critical component of any channel is the quality and type of programs they telecast. This determines their popularity, which in turn determines amount of advertisement revenues they can generate. They can do any one of the following:

Buy programming rights of program software from outside and collect advertisement revenue on their own. This model is followed by several TV companies, wherein they have a separate company in their fold, which develops all the content. The advantage is that re runs of serials/ programs become very profitable.

Selling time space to the producers for a fixed charge. Producers in turn are free to book advertisements at their own rates (there is an understanding on the time allocated for advertisement) and collect revenue. This is the basic model for many TV companies in which they sell prime time slots. The rights continue to be vested with the producer.

Earnings drives

The key factors that drive sector revenues are

Television penetration: Since the medium is television, increased television penetration will imply higher viewer ship. This will translate into higher advertisement spend allocation. This will also imply higher software production and demand for new programs.

Competition from other satellite channels would have an adverse impact on advertisement revenues, as advertisers have more choice in allocating ad budgets.

Government policies can have a big impact on the fortunes of the entire industry. When the DTH bill is passed in any country then, it will trigger a restructuring of the cable business.

Launching new channels targeted at specific segments, like regional channels within any country other areas having large pockets of ethnic population would lead to revenue growth. This will entail significant initial outlays.

Depreciation of the local currency would increase revenues as most of the program/ software companies export the programs overseas and payments are dollar denominated.

Advertisement revenue

As mentioned earlier, this is the primary source of income for TV channel operators. This revenue is directly co-related with the reach and viewer ship of a channel. Any channel's popularity depends on good quality programs, which is the software content. The business requires enormous initial investment in programs and revenues follow only with a time lag after the channel receives a minimum viewer acceptance.

Outlook

The sector has latent potential for growth on back of the exponential growth of cable TVs during the last 5 years. Television penetration in Africa is extremely low as compared to other developing countries like Malaysia, Pakistan, etc. in Asia. The number of channels has increased, implying higher demand for software programs.

Advertisement revenues, which are the barometer of channel popularity, will get dispersed over several competing channels. A shakeout is likely in both the channel and cable TV sectors. The biggest beneficiaries will be the content providers or the software houses. They will control the intellectual rights to the key element driving any channel's popularity.

Direct-to-Home, Digital Terrestrial Transmission and Conditional Access Cable Delivery have emerged as new delivery mechanisms. Breakthrough in technology would help open up avenues for these channels.

ANGOLA TELEVISION NETWORK

MARKETING

CHAPTER - 7

MARKETING

7.1   Description of the commercial viability of the project with regards to revenue generations.

Estimations shows that, after about 5 years of regular working, the CPLP Cultural Television Channel will be able to start exporting a great quantity of services and videos to Africa, Europe and South America. (See financial projections)

The economic feasibility of the project will occur with the commercialisation of several products, like:

1.	The sale and production of cultural pro-
grammes to various institutions like:
-	BBC
-	TV Culture Brazil
-	RTP
-	Discovery
Etc.
All over the world commercial networks for television cultural programmes have been created.
Several television channels in different countries
have regularly acquired programmes by producers spread out all over the world. Not only, several
international organisations, like the World Bank
and the United Nations, including FAO and Unesco,
among many others, need programmes for public
education, like programmes oriented to alert and
to educate people concerning endemic and epidemic diseases. AIDS, malaria, tuberculosis are a
very few examples.

These world institutions have made a great effort
to develop television programmes with humanitarian objectives. But, in Africa – undoubtedly the continent most needed of such programmes – there is
no television station, in present times, with capability to make face to such a need.

Therefore, the television teams that had been responsible for such programmes are, practically in

its totality, placed in countries of the called First World, strange to the local population's concrete reality.

Programmes focusing new agricultural techniques or even oriented to agricultural, commercial and industrial education are essential elements of such a repressed demand.

The price for each institutional campaign, with an average of 10 films produced per campaign, is of about USD.15.000$ and the capacity of the CPLP Cultural Channel will be as follows:

First Group* - from 2 to 4 campaigns per month.
Second Group – from 1 campaign in two months to 1 campaign per month

This represents:
First Group   a potential annual income of about US$.540.000
Second Group a potential annual income of about US$.90.000

*First Group: Angola, Cape Verde, and Mozambique. Second Group: Guinea-Bissau, Sao Tome e Principe, East Timor (Lorosae). Reference Group: Brazil, Portugal.

2. The sale and production of no cultural pro-

grammes to other television channels, like:
-       series (novels)
-       talk-shows Etc.

Many countries of the region need to produce television programmes, but do not have capability to do it. Thus, they seem themselves obliged to search expensive productions in Europe. With only one programme sold-three daily hours-the channels of the First Group* would receive in incomes the equivalent of about two times of the whole investment. The Second Group countries will not have conditions at the beginning to product cultural programmes to other television channels.

*First Group: Angola, Cape Verde, and Mozambique. Second Grope: Guinea-Bissau, Sao Tome e Principe, East Timor (Lorosae). Reference Group: Brazil, Portugal.

The price for this type of television programme is:

Auditory programs   US$.12.500 per programme
Talk-shows   US$.4.000 per programme
Interviews   US$.4.000 per programme

The capacity of production, of these programmes by the countries of the First Group will be:

In the FIRST PHASE:

Talk-shows   2 per month
Interviews   4 per month
In the FINAL PHASE:
Auditory programmes   6 per month
Talk-shows   20 per month
Interviews   20 per month

Thus, the potential income will be:

FIRST PHASE
Talk-shows   US$.96.000 per year
Interviews   US$.192.000 per year

FINAL PHASE
Auditory programmes   US$.900.000 per year
Talk-shows   US$.960.000 per year
Interviews   US$.960.000 per year

The price of the novels is much higher. For each complete novel, with about 120 chapters, the price is about US$.2.500.000 and the capacity of the First Group of the CPLP Cultural Channel (in the final phase) will be of one novel per year.

3. There is the possibility of commercial use of the transmission time beyond the six hours reserved to culture and education.

The conventional day in television is of 18 hours, of which only 6 hours would be no commercial. We would have, therefore, 12 hours of no cultural transmission, which should be freely commercialised.

Even the period of six hours- divided into two sections of three hours, the first one dedicated to culture and second section of three hours to education-has a great potential for sponsoring.

Each commercial hour of transmission can include up to 12 minutes of advertising. The price for each 30 seconds of advertising is:

Noble time   from 6PM to 11PM   a b o u t US$.400 each 30"
Normal time   rest   about US$.200 each 30"

Concentrating the commercial programming in the noble time. With 4 hours of transmission in this period, we would have:

Educational television: from 7AM to 9AM
From 3PM to 4PM
Cultural television: from 4PM to 7PM
Commercial television: from 9AM to 3PM
From 7PM to 1AM

Thus, the commercial period would comprehend 6 hours in normal time, 4 hours in noble time and 2 hours in normal time again.

Begin 12 minutes per hour for advertising, the commercial period could have:

Noble time per day	48 minutes	24	films	US$.9.600
Normal time per day	96 minutes	48	films	US$.9.600
Total of the potential income...........				US$.19.200

per day

> or US$.576.000 per month
> or US$.6.912.000 per year

It is believed that in the first phase of the project (after one year), the CPLP Cultural Television Channels of the First Group will be able to start with an income from commercial advertisement of about US$.1.000.000 per year-value which is predicted to increase in the follow months.

4. The rent of the studios to television teams of other countries can be another source of incomes. Many producers who cover events in Africa need to move many times to their countries of origin during the video works, because there is not technical support in Africa. The same phenomenon happens with the cinematography and the journalism productions. It is not difficult to imagine, for example, the serious problems journalism teams have had, for example, with essential components like batteries, lighting, electronic components etc., which only can be easily find in Europe.

The period of renting of a television studio is of 12 hours. Each period has price of about US.4.000. After the first year, the studio should be rented for 12 hours per each period of 3 days. So, the rent per month, in this period (First Phase), will be able to generate incomes of about US$.120.000 per month in the First Group.

5. Support services to other television networks in all areas, including novels, mini series, docdramas series etc.

Many countries, principally in Africa, do not have technical conditions to develop this kind of programmes, but they have a strong internal repressed demand in this sense.

Each team, abroad, has a price of US$.1.000 per day (two people), after the costs of dislocation, hotel and meals. In the first months (first phase) it will be organised one team for works in other countries. In the final phase it is predicted to have up to 5 teams with such function. The capacity of each team is of about 20 days per month. Therefore, the support to other television networks-when took in its full potentiality-will be able to represent up to 20 days per month in the first phase. This represents a potential income of about US$.240.000 per year, and 100 days per month of works in the final phase, signifying about US$.1.200.000 per year of incomes, always referring to the First Group countries.

6. Colloquies, seminars, videoconferences and meetings of different natures.

The building of the CPLP Cultural Television Network will have a medium size auditorium, (First Group), with all conditions to receive seminars, colloquies and meetings of the most diverse nature. Such seminars, videoconferences etc. are important not only for the increase of the incomes, but also to attract specialists of the most different areas, being an important element for the development of all region as well as for the diffusion of the local, regional and continental cultural values.

The meetings, seminars and colloquies attract, in average, about 300 people per event. The price-excepting meals, hotels and transportation-per each participant is of about US$.50 per day. It will be a capacity for up to four events of this type per month, what could represent an income of about US$.60.000 per month or US$.720.000 per year (final phase-after 5 to 7 years).

The project should also turn possible.
-		classes of professional formation in diverse disciplines
-		support for a multimedia high technology centre (First Group)
-		support for a multimedia high cultural centre (Final Group)

ANGOLA TELEVISION NETWORK

PROFITABILITY & CASH FLOW

CHAPTER - 8

PROFITABILITY AND CASH FLOW

8.1　Estimation of cost of production and working results for the first five years of operation.

The estimated cost of production of working results for the first five years of operation are given in the chapter "Financial Projection".

8.2　Cash flow statement for the company as a whole, for five operating years of the project based on the estimates of working results.

A detailed cash flow statement for the company as a whole for five operating years is given in the chapter of "Financial Projection".

8.3   Projected balance sheet for five operating years for the company as a whole.

The balance sheet for five operation years for the company as a whole is given in the chapter of "Financial Projection".

ANGOLA TELEVISION NETWORK

ASSUMPTIONS

CHAPTER - 9

ASSUMPTIONS

1.   The Angola Television Network shall generate the income from the following sources.

a)   Sale of cultural programmes.
b)   Sale of commercial programmes.
c)   Sale of advertisement time during transmission of TV programmes

d)   Hiring of studios.
e)   Supply of technical services
f)   Hiring of conference hall

The detailed assumptions for each of the above mentioned activities are as follows:

a)   Sale of cultural programs:

I.   It is assumed that four cultural programs shall be produced per month in the first year, which can be sold to other countries. This

figure will increase to 5 programs per month in the second year, six programs per month in the third year and so on.

II.   It is assumed that the selling price shall be US$.150 per program.

b)   Sale of commercial programs:

I.   The commercial programs consist of talk shows, interviews and auditory programs.

II.   It is assumed that two talk shows shall be produced per month in the first year, three talk shows shall be produced in the third year, four talk shows per month shall be produced in the third year and so on.

III.   It is assumed that the selling price of one talk show program shall be US$.4000.

IV.   It is assumed that four interviews shall be produced per month in the first year, five interviews shall be produced in the third year, six interviews per month shall be produced in the third year and so on.

V.   It is assumed that the selling price of one interviews program shall be US$.4000.

VI.   It is assumed that no auditory programs shall be produced in the first and the second year. Only when the people have two years of experience then in the third year one auditory program shall be produced per month. In the fourth year two programs shall be produced per month and so on.

VII.   It is assumed that the selling price of one auditory program shall be UD$.12.500.

c) Sale of advertisement time during the transmission of T.V. programs:

I.   There shall be a total transmission of 8 hours per day in the first year. It will go on increasing by two hours in the second and subsequent years.

II.   Out of which advertisement shall be available for programs with transmission period of four hours in the first year. The programs in

which advertisements could be available shall increase by two hours per year in the second and subsequent years.

III.   In one hour T.V. transmission, 12 minutes of advertisement shall be allowed.

IV.   Out of total T.V. time 33 % time shall be considered as prime time and remaining 67% shall be considered as non prime time, in the first, second and third years. From the fourth year onwards the prime-time advertisements shall remain constant.

V.   The advertisement rates shall be US$ 800 for one minute of prime time advertisement and US$ 350 for one minute of non prime time advertisement.

d)   Hiring of studios:

I.   It is assumed that the studios shall be taken on hire for a shift of 12 hours, two times in a month in the first year. It will increase to

three times in a month in the second year.
II.   The rent per day shall be US$.4.000.

e)   Supply of technical services:-

I.   It is assumed that one team, consisting of two technically qualified people shall be available in the first year. In the second year two such teams shall be available. In the third year three such teams shall be available and so on.
II.   It is assumed that in the first year the team shall be hired for 10 days in a month. In the second year the teams shall be hired for 12

days in a month. In the third year the teams shall be hired for 14 days in a month and so on.

III.   The rate of hire shall be US$.1.000 per team per day.

f)   Hiring of conference hall :-

I.   It is assumed that the conference hall shall be taken on hire for two times in a month in the first year, three times in a month in the

second year and four times in a month in the third year and so on.

II.   The hire charges shall be US$.500 per conference.

2.   The unit is planning transmission of three hours of educational programs, three hours of cultural programs and two hours of commercial programs per day in the first year. Thus, there shall be a total transmission of eight hours of programs per day in the first year. The total transmission hours shall increase by two hours per day in the second and subsequent years. Thus, in the second year there shall be total transmission of 10 hours per day and in the third year there shall be a total transmission of 12 hours per day. The cost of software production is assumed at US$.2.500 per hour.

3.   The cost of miscellaneous consumable

items is assumed at US$.36.000 in the first year. This will increase by 10% in the second and subsequent years.

4.   The cost of stores and spares consumed is assumed at US$.42.000 in the first year. It will increase by 10% in the second and subsequent years.

5.   The cost of repairs and maintenance is assumed at 0.1% of the total cost of plant and machinery in the first year. This cost will increase up to 0,3% in the fifth year.

6.   It is assumed that the unit shall use 1.000 units of power every day or 365.000 units of power in the first year. The consumption of power shall increase by 10% in the second and subsequent years. The cost of power is assumed at 5 cents for one unit.

7.   It is assumed that the unit shall consume 1.000 litters of diesel per month. The cost is assumed at 33 cents for one litter. It will increase by 10% in second and subsequent years.

8.   The unit shall have a technical staff of 35 people and administrative staff of 15 people.

9.   Depreciation is calculated on reducing bal-

ance method. The rate of depreciation is taken at 10% for building, 25% for plant and machinery, 10% for furniture and fixtures and 20% on vehicles.

10.   Advertisement expenditure is assumed at 0.2% of the sales.

11.   It is assumed that the stock of raw material and stores shall be 30 days of consumption, receivables shall be 45 days sales and creditors shall be 15 days purchases.

12.   The term loan shall be for a total period of 25 years. It shall have a moratorium of five years. It shall be repayable in 20 equal yearly instalments.

PROJECT REPORT

OF

CAPE VERDE CULTURAL TELEVISION NET-WORK

AT

CAPE VERDE

AFRICA

PROMOTERS

REPUBLIC OF CAPE VERDE

AFRICA

TELEVISÃO SONHO

Rede Planetária de Televisão Cultural para a Paz Mundial

CAPE VERDE TELEVISION NETWORK

INTRODUCTION

CHAPTER - 1

INTRODUCTION

1.1   Description of the organisation through which this project is being taken up.

The CPLP is an international organisation, created in 17 of July of 1996, with Headquarters in Lisbon and consisting of the following State Members: Angola, Brazil, Cape-Verde, Guinea-Bissau, Mozambique, Portugal, Sao Tome e Principe and East Timor (Timor Lorosae).

The CPLP has as its objective the politic-diplomatic relationships between its Members, mainly referring to the United Nations and the World Bank, the relationship EU/MERCOSUL and the accomplishment of the Europe-Africa Summit.

The CPLP is also dedicated to the co-operation, particularly in the economic, social, cultural, legal and technical-scientific matters; as well as to the projects of promotion and broadcasting of the Portuguese Language, nominated to the improvement of the International Institute of the Portuguese Language and the creation of a Bibliographical Fund.

All its decisions are taken by consensus.

2.2   Description of the main reasons for setting up an African Cultural Television Network.

Africa is a continent submerged into dramatic problems. All these humanitarian problems have in the education and culture their most direct and objective root.

The African countries speaking Portuguese are among the African countries with the lowest level of development.

The whole situation is so dramatic that the implementation of the CPLP Cultural Television Network is more than urgent.

It is hoped that with this Television Network the people shall have better understanding of various diseases and other problems of life. e.g. AIDS is prevalent disease in Africa. But as the people are illiterate the information about the disease cannot be spread by any other method. In such cases the Television Network becomes a very effective audio-visual means of communication.

The project also intends to be a base of a massive production of health campaigns, specially to reinforce the combat against the AIDS, the infect-contagious diseases- endemic or epidemic-through a process of global education of the local and regional populations.

The CPLP Cultural Television Network is a project for peace and its objective is not to compete with local television channels already installed, but to interchange with them, promoting a future global web dedicated to knowledge and development.

1.03 Description of the main objects of the project.

The major objective of the project is to assist the CPLP countries to develop the first African television network channel oriented to cultural and educational themes with the follow orientation:

-	diffusion of the Portuguese language
-	education in all areas
-	reinforcement of the local and regional cultures
-	formation of technical personal on communication and tele-education
-	formation and information for a better participation of the CPLP nations in the process of global development

- to support the implantation of the Institute of Portuguese Language

The development of such a project will create an irradiation pole from the countries with the project implementation to the whole African continent, constituting an African network of research and culture, discovering and valorising different cultural aspects and, consequently, reinforcing identity factors.

With the stronger cultural identity, it is expected to have lower levels of violence in social behaviour as well as a relevant increase of the educational performance in all its sectors.

Systematic educational objectives are complementary to the general orientation of this project, which is cultural in its fundament.
A very large positive impact is also to be expected with respect to several sectors of the economy, greatly improving the overall quality of services.

1.4   Description of the mission status of the project.

The mission of the project is:

- to rescue lost values and cultural information
- to reinforce historical elements

- to rescue old artworks
- to give a new value to new artworks
- to communicate practical and objective information on health
- to communicate practical and objective information on agriculture
- to communicate practical and objective information on economy

1.05 Description of the philosophy of the project.

The CPLP Cultural Television Network will have in its philosophy, as a support to its mission, the follow fundaments:

- to work with local people
- be focused on quality
- intercommunication

Three words can define the CPLP Cultural Television Network philosophy:
People, quality and intercommunication

Following these three key words of its philosophy, a single sentence also defines its mission: "A Continent of Culture".

Two main sectors characterise its global structure:
1. Cultural Sector
1.1.   To develop programmes on:
- Music

- Architecture (also vernacular)
- Archaeology
- Literature and poetry
- History
- Tourism
- Dance
- Plastic arts
- Astronomy
- Science in general
- gastronomy
- health
- anthropology
- local cultures
- others

1.2   The programmes can be of all types, the most important is to be attentive to communication. All programmes must reach the highest possible number of spectators.

2. Educational sector.
2.1   Systematic education in short practical courses.
2.2   Educational series with practical and objective information on:
- health care
- family
- agriculture
- managing
- services
- civil construction

As the CPLP Cultural Television Network must not be supported by the respective governments in the future, it will have a daily time of free commercial programmes up to 12 hours. This is a good solution to pay its costs and support its development.

1.06 Description the magnitude of the project.

The project proposes a Television transmission time of eight hours per day in the first year. It will increase by two hours per day in the second and subsequent years.

The project also envisages training of local people in all the related areas of Television. It is proposed to develop a local pool of talent which can be utilised not only in this particular country but also in other countries of Africa.

The total cost of the project of the Cape Verde Cultural Television Network is US$.18.400.000. Out of this it is proposed to obtain a soft loan of US$.15.400.000. The remaining amount of US$.3.000.000 shall be contributed by the Government of Cape Verde in terrain for the installation of the respective building, infrastructures and logistic support.

For the purpose of the Cape Verde Cultural Television Network a company shall be formed with the participation of the Government. The Contribution of the Government in the form of share capital shall be US$.3.000.000.

The term loan shall be on soft terms. The total period of the loan shall be 23 years. Out of which five years shall be a period of moratorium. The entire loan shall be repaid in the remaining 17 years, in 17 equal yearly instalments.

The Government of Cape Verde will give the sovereign guarantee for the referred loan.

CAPE VERDE TELEVISION NETWORK

THE COUNTRY

CHAPTER - 2

THE COUNTRY

2.1 - Description of the level of development of Cape Verde.

Cape Verde's low per capita GDP reflects a poor natural resource base, serious water shortages exacerbated by cycles of long-term drought, and a high birth rate.

The economy is service-oriented, with commerce, transport, and public services accounting for almost 70% of GDP.

Although nearly 70% of the population lives in rural areas, the share of agriculture in GDP in 1995 was only 8%, of which fishing accounts for 1.5%. About 90% of food must be imported. The fishing potential, mostly lobster and tuna, is not fully exploited. Cape Verde annually runs a high trade deficit, financed by foreign aid and remittances from emigrants; remittances constitute a supplement to GDP of more than 20%. Economic reforms, launched by the new democratic government in 1991, are aimed at developing the private sector and attracting foreign investment to diversify the economy. Prospects for 1998 depend heavily on the maintenance of aid flows, remittances, and the momentum of the government's development program.

OFFICIAL NAME: Republic of Cape Verde
CAPITAL: Praia
SYSTEM OF GOVERNMENT: Multiparty Republic
AREA: 4,033 Sq Km (1,557 Sq Mi)
ESTIMATED 2000 POPULATION: 437,500

LOCATION & GEOGRAPHY : Cape Verde is an island group or archipelago of ten islands and five islets located in the Atlantic Ocean off the west coast of Africa. The archipelago is divided into two groups (1.) the Barlayento or windward islands in the north and (2.) the Sotavento or leeward islands in the south. The islands are of volcanic origin and except for the Boa Vista, Maio and Sal are mountainous with rugged cliffs and steep ravines.

The coastal plains are semi desert with fine sandy beaches while the mountains are covered by thin forests. Major Cities (pop. est.); Praia 61,700, Mindelo 47,100, Sao Filipe 5,600 (1990). Land Use; forested 1%, pastures 6%, agricultural-cultivated

11%, other 82% (1993).

CLIMATE: Cape Verde has a tropical climate with two seasons. A cool dry season from December to June and a warm season between July and November. Rainfall is low and unreliable with most of it occurring during August and September. The islands suffer from severe shortages of water and rainfall which cause catastrophic and prolonged droughts periodically. Tropical heat and high humidity prevail throughout the year and the conditions are uncomfortable except when fanned by the northeast sea breezes. Average temperature ranges in Praia are from 19 to 25 degrees Celsius (66 to 77 degrees Fahrenheit) in February or March to 24 to

29 degrees Celsius (75 to 84 degrees Fahrenheit) in October.

PEOPLE: Around 71% of the population are Creoles of mixed Black African and Portuguese descent while the remainder are almost all Black Africans with a small number of Whites.

DEMOGRAPHIC/VITAL STATISTICS: Density; 85 persons per sq km (219 persons per sq mi) (1991). Urban-Rural; 29.7% urban, 70.3% rural (1990). Sex Distribution; 48.0% male, 52.0% female (1990). Life Expectancy at Birth; 60.0 years male, 64.0 years female (1992). Age Breakdown; 45% under 15, 31% 15 to 29, 10% 30 to 44, 7% 45 to 59, 7% 60 and over (1990). Birth Rate; 48.0 per 1,000 (1992). Death Rate; 10.0 per 1,000 (1992). Increase Rate; 38.0 per 1,000 (1992). Infant Mortality Rate; 61.0 per 1,000 live births (1992).

RELIGIONS: Mostly Christians with the Creoles and the majority of the Black Africans of the Roman Catholic faith which accounts for around 98% of the population.

LANGUAGES: The official language is Portuguese, although the national language is a Portuguese Creole known as Crioulo.

EDUCATION: Aged 25 or over and having attained: no formal schooling or incomplete primary 84.2%, primary 12.4%, secondary 1.7%, higher 0.5%, not specified 1.2% (1980). Literacy; literate population aged 15 or over 73,500 or 47.4% (1985).

CURRENCY: The official currency is the Escudo (CVEsc) divided into 100 Centavos.

ECONOMY: Gross National Product; USD $347,000,000(1993). Public Debt; USD $148,000,000 (1993). Imports; CVEsc 12,387,000,000 (1993). Exports; CVEsc 312,200,000 (1993). Tourism Receipts; N/A. Balance of Trade; CVEsc -12,075,000,000 (1993) . Economically Active Population; 120,565 or 35.3% of total population (1990). Unemployed; 25.8% (1990).

MAIN TRADING PARTNERS: Its main trading partners are Portugal, Spain, France and the Netherlands.

MAIN PRIMARY PRODUCTS: Bananas, Basalt, Beans, Coffee, Kaolin, Livestock, Maize, Pozzuolana, Salt, Sugar, Sweet Potatoes, Yams.

MAJOR INDUSTRIES: Agriculture, Canning, Cement, Fishing, Food

Processing.

MAIN EXPORTS: Bananas, Coffee, Fish, Salt.

TRANSPORT: Railroads; nil. Roads; length 5,615 km (3,489 mi) (1987). Vehicles; cars 13,027 (1988), trucks and buses 4,356 (1988). Merchant Marine; vessels 40 (1990), dead weight tonnage 29,730 (1990). Air Transport; passenger-km 122,959,000 (76,403,000 passenger-mi) (1987), cargo ton-km 2,345,000 (1,606,000 short ton-mi) (1987).

COMMUNICATIONS: Daily Newspapers; nil. Radio; receivers 57,000 (1994). Television; receivers 5,000 (1987). Telephones; units 15,000 (1994).

MILITARY: 1,100 (1994) total active duty personnel with 90.9% army, 0.0% navy and 9.1% air force while military expenditure accounts for 1.0% (1992) of the Gross National Product (GNP).

CAPE VERDE TELEVISION NETWORK

THE PROJECT

CHAPTER - 3

THE PROJECT

3.1   Description of the initial steps to be taken for the implementation of the project.

The first objective of the CPLP Cultural Television Network is to install its structures in the following countries:

- Sao Tome e Principe
- Angola
- Mozambique
- Guinea-Bissau
- Cape Verde
- Timor

Thus, these six countries will structure the basic network of the whole project.

Cape Verde, Mozambique and Angola will form a FIRST GROUP; Sao Tome e Principe, Guinea-Bissau and East Timor will form a SECOND GROUP; and, finally, Brazil and Portugal will form a called REFERENCE GROUP.

However, the project will start at the same time in all countries. A more precise description of the meaning of each group is made below.

The basic physical components of the project are:

1. The creation of companies, in each of the States of the CPLP, oriented to the objectives above referred.
2. The State of each country – who will be the responsible and guarantee of the payment of the financing – will also have a participation in the companies.
3. All strategies will be oriented to lowest costs in long term.
4. The whole project will be developed by phases.
5. The timetable of transmission will be defined case by case.
6. The project will include:
6.1. a complete plan for an equipment network.
6.2. a complete plan for a software network.
6.3. all procurements
6.4. equipment transportation and installation
6.5. all configurations
6.6. creation of the infographic system
6.7. technical formation of the personal
6.8. architecture and engineer projects for the building

6.9.   construction of the building organised in phases

1.2   Description of the procedure by which the necessary programmes will be produced for transmission on the channel.

The television programs are known as software in this field. The unit is planning transmission of three hours of educational programs, three hours of cultural programs and two hours of commercial programs per day in the first year. Thus, there shall be a total transmission of eight hours of programs per day in the first year. The total transmission hours shall increase by two hours per day in the second and subsequent years. Thus, in the second year there shall be total transmission of 10 hours per day and in the third year there shall be a total transmission of 12 hours per day. The cost of software production is assumed at US$.2.500 per hour. The amount of US$.2.500 per hour is taken based on cost of production of such programmes in Portugal.

To contract a similar system from other television channel, only for one country, for a transmission period of 6 hours daily, the cost in, market (eg. Portugal*) is of about US$.500.000 per month. This cost represents a total of about US$.6.000.000 per year. In 16 years, excluding the first year of implementation, the costs for contract the six daily hours of transmission from other television channel, in present market price, is of about US$.100.000.000. The six countries would have an estimated cost of about US$.600.000.000 in a 16 years period of time.

These costs are based on the prices presently used in Portugal, as for example the programme Hora Viva – Seguranca em Directo, transmitted every day, excluding weekends, from 7 to 10 AM.

Another example is the channel Canal Noticias Lisboa CNL. Only in its first year of installation, it was expended about US$.10.000.000, and the equipment as well as the building were not property of the channel, but rented.

In fact, however the project will be implemented by phases, after 5 to 10 years of regular work, it is predicted to have reached a similar value in incomes as the examples above.

*value took in the Portuguese market, below the average of the European prices in about 30%.

The first ten months of works will be oriented to:
-        design of the equipment and software structures
-        start the first productions
-        elaboration of the projects of architecture and engineering

-        procurement and acquisition of all equipment
-        construction and supervision of the works
-        installation and configuration of the equipment

So, the first year must be used to the implantation of the project. However, as to accelerate the chronogram, some operations should be started before this timing, temporarily installed in a different building.

The first phase will be oriented to:
-        video productions (external and internal)
-        edition
-        elaboration of cost programmes

3.3   Description of the requirements of Power, Fuel and Water. POWER
It is assumed that the unit shall use one thousand units of power of everyday. Thus, the unit shall use 365.000 units of power in a year. The costs of power is assumed at 3.5 cents for one unit. The consumption of power shall increase by 10% every year.

FUEL

The unit shall use D.G. set for generation of electricity, when there is an interruption in the supply. The hours of disruption in the normal supply can not be predicted accurately. Hence, the hours for which the D.G. sets shall be used also can not be predicted accurately. However, it is assumed that the unit shall utilise one thousand litters of diesel per month. The cost is assumed that 23 cents for one litter. Hence, the total cost for one thousand litters shall be US$.230. The consumption of diesel shall increase by 10% every year.

WATER

The unit does not need water for fixed commercial operation. However, it shall have a strength of 50 people as employees. Water shall be required for drinking and sanitation purposes. It is assumed that 3000 litters of water shall be required everyday. The cost of obtaining this water is assumed at US$.7 everyday. The consumption of water shall increase by 10% every year.

3.4   Description of the manpower requirement of the project and description of the steps taken by the company to train the manpower.

The total staff strength shall be 17 people in the first year.

The system of modern TV Station is designed to achieve sustained operating efficiency and transmission. This, of course, entails a certain degree of sophistication in the production and transmission. However, considering the socio-economic situation in Cape Verde, a reasonable balance has to be struck to obtain optimum performance and at the same time create gainful employment. While working out the manpower requirement for this project to be kept on direct rolls of the company totalling 50 staff and operators, the above consideration have been kept in mind.

Manpower requirement

The direct manpower required for the proposed Unit is about US$.1.260.000 per year. The manpower requirement as indicated in this chapter has been planned keeping in view the following guidelines:

Effective co-ordination among the various departments. Judicious distribution of responsibilities. Capacity utilisation of the TV Station with optimum manpower.

Details of manpower requirement for the TV Station is given in the financial section.

In line with the prevailing practice, the Security

guards, office peons and unskilled labour etc., are normally employed on contractual/casual basis, and their cost has to be included in the Factory over-heads. However in this report provision has been made to employ the above staff and their salaried have been included.

The various departments proposed shall be under the direct responsibility of the Station Director. He shall be assisted by Deputy Directors, who will look after the complete TV Station and its various day to day activities and Marketing. They shall be respon-sible for achieving the envisaged targets and sales forecasts. They shall be assisted by a team of Man-agers from production, marketing and accounts.

To run this project a labour force which shall usu-ally be composed of unskilled and skilled workers shall be employed. The first are those who do not undergo any king of specific training or education, while the latter have to do so in order to master their jobs.

When evaluating an investment project from an em-ployment point of view, its impact on both unskilled and skilled labour has been taken into account. Not only direct employment, but also indirect employ-ment has been considered. Direct employment re-fers to the new employment opportunities created within the project; indirect employment concerns job opportunities created in other projects linked with the project which is being formulated.

The implementation of large and sophisticated projects generally contributes to the development of local skills and capabilities in a country. Furthermore, they help to change traditional values, attitudes and the behaviour of the society, to build up an enterprising spirit among the people, to develop a desire for changing and improving the existing conditions of life, to introduce better work discipline and thus to change the very pattern and basis of economic development. The TV industry is already well established in Africa. Location of TV industry activities in Africa, has certain favourable factors and advantages, in setting up this type of industry, as availability of acceptable level of education among the supervisory staff adequate technical and managerial skills developed over a long period, and availability of cheap labour are assured.

The organisation structure will vary from TV Station to TV Station in the industry and as such the pattern proposed herein can be considered only as suggestive and provisional.

A well-knit organisation structure headed by a Station Director and Manager, with the supporting staff will be developed progressively during project implementation. Soon after the plant becomes operative, a good number of project staff will be absorbed in the organisation. Certain additional staff also get added to ensure smooth and efficient management of the operating unit.

The Manager Production will have a degree in communication and will be accountable and answerable to the Station Director for all TV operations including planning, production, material management, TV utilities, quality control, production cost and budgetary control, TV safety, discipline and layout relations. His main function should be to ensure achievement for quality and quantity targets of production at reasonable cost and should constantly strive to improve TV performance. In the discharge of his multifarious duties and responsibilities he will be assisted by supervisors and adequate staff for day to day activities.

His duties are to ensure that targets are maintained through effective utilisation of 3 M's viz: Machinery, Men and Materials. These targets should be translated in terms of targets for individuals under him such as supervisors, skilled workers etc. He has to ensure that the equipment at his disposal gets prompt attention on breakdown is adhered to strictly. He has to further ensure that there is a strict quality control exercised over raw materials also. He will be assisted by Deputy Manager etc. who and will report to the General Manager.

GENERAL MANAGER (Comm.) He will report to the Station Director. His main duties will be prepare sales forecasts and budgets, study and monitor the export market for the company's products and advice on ways and means of increasing sales. He will also ensure that consumer complains are solved in an appropriate manner. He will also ensure that the right materials are available for production at the right time.

He will be responsible for all aspects connected with the export procedures and keep the management updated on all matters relating to the Govt. policies on polymers and Exports and above all the world market.

COMPANY SECRETARY & FINANCIAL CONTROLLER. He will report to the Station Director and will be responsible for the departments of administration, financial planning, budgetary control, cost accounting, tax management payroll accounting, etc. and all personnel matters. He will be assisted in their duties by their respective assistants to assist in day to day activities.

Manpower planning and production

The central issue here may be one of scale. Production is staffed bye personnel, but the process may be labour-intensive or capital-intensive. Either way, planning will include:

1.   Analysis of labour supply and demand factors in relation to skills and training needs.
2.   Procedure for manpower recruitment, together with selection processes.
3.   The formation of industrial relations policies necessary for effective work place bargaining, disciplinary measures and dismissal procedures.
4.   Analysis of the effective use of human resources.
5.   Conditions necessary to maintain adequate levels of motivation. The scale of the problem is likely to be directly proportional to the method of production.

The availability of main persons is not going to be easy since TV industry is currently in its infancy in Africa.

Training needs

The selection and training of the required manpower for the proposed project has to be planned in advance.

The key personnel should be selected and trained suitably. The training would be carried out in the following manner:

Basic training on the concept of TV industry before

construction begins with visits to similar TV Stations in other countries. On site training during the construction phase of the project. On job training during the commissioning phase of the project. On job training during operation of the TV immediately after commissioning. The training of the key personnel such as Station Director should be carried out in all the phases. The training of other operating personnel should be suitably carried out during the construction, erection and operation phases in addition to training them by visits to similar plants operating in other countries.

Besides training the key operating staff described above, in TV training should also be given to other employees at skilled operating level to enable them to understand the process equipment in the project and prepare them to operate a maintain their respective sections safely, efficiently and skilfully. The above training should be carried out during construction, commissioning and operating phases of the project.

Training is necessary in order to enable personnel to acquire the skills and knowledge necessary to perform a task to an acceptable standard. The length of the training period and training methods will, of course, vary from job to job. Training is essentially a learning process, and in order that progress can be successfully monitored certain conditions are necessary.

1.   The training needs of both the individual and the organisation shall be identified and analysed.

2.   Targets and standards shall be set for the trainee, which are within his capabilities.

3.   The pace of the training programme should reflect the trainee's ability to maintain progress in properly absorbing the same.

4.   The trainee shall receive regular feedback of results. Any problem areas shall be highlighted, discussed and resolved.

5.   As the trainee progresses the amount of information provided shall be gradually reduced, thus inducing a feeling of independence and competence.

It is common place to find a wide variety of tasks in an organisation and each will require varying degrees of skill, effort and responsibility. This being so, it is inevitable that rates of pay will also vary and the differentials between the jobs will reflect their relative values. However, other factors such as local market conditions, bargaining strengths and traditions also influence a company's payment structure and a great deal of planning is required if rationalisation is to be achieved. One technique which has been successfully adopted by many companies to establish an equitable wage structure is job evaluation. In a job evaluation exercise a comparison is made of common criteria over a range of jobs, and the resulting analysis may be linked to a points allocation or job ranking system, and hence to a wage scale.

In conducting a job evaluation exercise it is important to cover a reasonable variety of tasks within the whole spectrum. For each, a job description is prepared setting out details of the duties and responsibilities undertaken by the employee together with a statement about his working conditions. Very often this task is undertaken by work study personnel since, they are responsible for determining methods of operation and evaluating the work content of the job. Each job is assessed factor by factor, resulting in a comprehensive comparative analysis.

The individual is the most important resources of any company and only people who are well trained, well motivated and adequately rewarded will provide a positive and synergistic contribution towards the company's objective and its organisational health.

In most cases, the factors, which may be weighted according to relative value, are as follows:

1.   Skill-education, experience and training.
2.   Effort-both physical and mental.

3.   Responsibility-for equipment, materials, initiative etc.

4.   Working conditions-general conditions, risk of accident and injury.

Pay policies affect not only individual employees but the whole organisation, and the rewards and objectives vary at different levels within the enterprise.

Industrial relations

An effective industrial relations policy is important, since is the system through which employees take part in decision-making and in many instances it affects the while atmosphere of employer/employee relationships. An industrial relations policy is essentially a set of rules whose determine procedures for negotiation on such matters as:

1.   Wage and salary scales.
2.   Terms and conditions of work.
3.   Disputes and grievances.
4.   Recruitment and dismissal.
5.   Other issues of mutual interest, e.g. closed shop, redundancies and joint consultation.

In order to promote an atmosphere of co-operation, and to minimise conflict, the needs of management and work people must be recognised by both sides. Trade unions exit to protect the interests of their members and improve their working conditions. Management, while aware of the pressures and constraints imposed by the trade unions, have a duty to maximise the use of resources at their disposal, which may be expressed in relation to profitability, return on investment, level of service, sales volume, market share and cost-effectiveness. The strategies adopted in attempting to solve industrial relations problems will vary from company to company, and indeed from union to union, but there is no doubt that they will be influenced by both internal and external factors.

Internal factors

1.   The attitudes of employees to management, and management to employees.
2.   The leadership style of management.
3.   The bargaining strength of both parties.
4.   The number of negotiating bodies.
5.   The prosperity of the company.

External factors

1.   The extent to which parent boards influence company management, and district officials influence or control local shop stewards.
2.   Whether or not bargaining is conducted at plant, local or national level.

3.   Government policy towards industrial relations.

4.   The economic situations nationally, locally or within the company itself.

3.5   Describe the project implementation schedule.

Implementation of this project is a challenging task and calls for meticulous planning, scheduling and monitoring to realise the project goals within the budgeted cost and time frame. The goal can be achieved by adopting modern project management techniques.

To implement this project adequately, a team of engineers and project personnel having requisite education and experience are being appointed, to whom a detailed Work Breakdown Structure (WBS) in a logical order of activities, shall be supplied shortly, keeping in view cost estimation, scheduling, and to help monitor and control of the project. It is proposed to be formulated in conjunction with the objectives of each activity and goal settings. Project shall be programmed and controlled by network analysis techniques. Before the application of the network analysis techniques, the project personnel shall be acquainted with their capabilities in saving time, resources and costs. The training of project personnel and engineers at levels shall be provided for proper control of project progress and taking of timely corrective actions to re-align these efforts to meet ore stated objectives. It is proposed to gear programming and control system, i.e. project implementation system, which will ensure an integrated approach to project implementation. Project management activities shall be determined in advance and all activities carried out be project personnel as well as those to be contracted shall be identified.

Responsibility for project implementation shall be clearly defined. The forms of project organisation range from project oriented to functional organisation, while most of the cases are combinations of the two, with certain adaptations to prevailing conditions. It is impossible to over emphasise the importance of establishing a team of a task force for implementing the project with a designated leader to co-ordinate and guide its functions.

Project manager: Shall be responsible to the Board of Directors. The project manager shall be responsible for guiding and co-ordinating the efforts of all parties engaged in implementing the project, obtaining necessary government approvals on contracts. He is to control the project organisation with the promoters as well as with other agencies and organisations interested in the project. The manager shall have some staff to assist him, especially in checking expenditures to date and determining the present and future cost overrun or under run so that the project manager can take or propose to the Board pertinent corrective measures.

The network shall cover the pre-construction phase of the project indicating major administrative processes, since experience shows that some of them have frequently involved lengthy delays. In other words, it shall include the aggregate activities to be carried out be the principal parties participating in the implementation process.

The project budget shall then be prepared. Part or periodic payments to contractors which might be made at the end of certain intervals (e.g. weekly or monthly) throughout the time horizon of project implementation, shall be made by summing up activity costs per unit of time, which may be a week & month, and computing the cumulative cost at the end of each time interval. For the activities that are in process and are contacted or sub-contracted, the assumption of a linear time cost activity relationship shall be used for the sake of simplicity. In other words, expenditures are uniformly distributed throughout the duration of the activity.

PROJECT SCHEDULE

After an investment decision is taken, the main machinery and long delivery items must be ordered out at the earliest, forming the first major step in implementation of the project. It is foreseen that an engineering consultant will be appointed for carrying out the detailed engineering including basic engineering and procurement assistance to the client. It is also assumed that reputed and experienced contractors with adequate resources viz., men, materials, tools and tackles etc. will be engaged for execution of the construction and erection work. The purchase packages for auxiliaries shall be kept minimum so as to reduce the co-ordination efforts to the minimum. A great deal of co-ordination is required for constructing/erecting the new units. This task is feasible, provided the major activities of the project are co-ordinated and completed in the duration specified to achieve the respective milestones in time.

STRATEGY FOR TIMELY EXECUTION

It is important to deploy a team of experienced personnel for project execution and select the external agencies with due care for rendering the services and supply of equipment for the project. The project activities must be identified, planned and scheduled, and the progress monitored for timely project implementation. All the inputs to the project including financial resources must be identified and their inflow planned and arranged in time.

The project must be managed professionally with necessary co-ordination among the various agencies and requisite decisions taken promptly.

Establishment of an effective monitoring procedure for progress review and co-ordination.

However this tentative schedule, many activities will begin in the 6th month after the start of the financing, using different locations and training some partial personal. This will turn possible the strength of the dynamics in the whole process as well as the shorting of the estimated timing.

In short, the following key factors would constitute the broad strategy for timely execution of all activities in a pre-determined manner as per schedule shown in the bar chart as to reach at a basis of regular production.

Early selection of an effective in-house technical team (TASK FORCE) by Government of Cape Verde, headed by a Project Manager for planning and executing the project.

i)   Proper choice of external agencies such as consultants for Project Engineering., Machinery suppliers, Construction Agencies etc. keeping in view their reputation/past performance and working experience in their fields.

ii)   Adequate use of computer-based PERT/ CPM techniques for project planning, scheduling and monitoring.

3.6   Description of the initial project Flowchart.

The project implementation phase embraces the period from the decision to start the project to the beginning of the commercial production. It includes a number of stages including negotiations and contracting, project design, construction and start-up.

3.7   Description of the requirement of Land.

For this project the Government of Cape Verde will participate with about 25.000 square meters of land. In selecting the land the following criteria should be followed.

1.   The land should be near to the main city of Praia.
2.   The land should be properly connected by road.
3.   It should also have proximity to many end user clients.
4.   It should also meet various Government policy of

achieving the social objectives.

2.08 Description of the requirement of Building.

The total requirement of building is as follows.
a. Adm. Building   700 Square Meters
b. Studios –3 (Three)   1.800 Square Meters
c. Other Misc. Building   300 Square Meters

Total Square Meters: approximately 2800 Square Meters.
Cost for construction of about US$750 per square meter (including air conditioning systems, primary electrical cabin etc.).

2.9   Description of some requirements for Plant & Machinery and their basis of selection.

The plant and machinery have been selected giving due consideration to the sophisticated nature of technology required. Detailed discussions were carried out with the foreign suppliers to ensure that the required capacities are practical with minimum capital and operating costs before the machinery were finally selected.

The machinery shall be selected from the most reputed foreign manufacturers of complete range of equipment. Keeping in view the following main factors:

a)   Past performance
b)   Existing machinery in Africa & abroad.
c)   Sales and service facilities in Africa

The main equipment is as follows.

STUDIO

1.   Post-production equipment – non-linear post-production, two ES7 stations.
2.   ENG – 3 DVCAM Camcorders
3.   Régie with 4 DXC-D30PK1 Digital Cameras - for cold light installation.
4.   Flexicart.
5.   Computer graphics suite.
6.   Sound room.
7.   MUST SYSTEM: Automatic Emission Managing System, with graphical information for no active emissions. (This system has been used by several French thematic channels)

Other technical data:
The whole installation is characterised by:
1. Informational web with UTP level, 5 cables.
2. Phone system.
3. Specific furniture.
4. Parking for about 200 places

Discrimination of Technical Data:

Studio
Cold lighting system.
4 Camera channels composed by:
2 DXC-D30PK Sony Digital Cameras
4 J18 Canon Zoom Lens
4   CA-537P Sony Camera Adapters
4   DXF-50 Sony Studio Viewfinder's
4   CCU-M3P Sony Control Camera Units
4   RM-7P Sony remote CCU controls
4   50M Sony Multicore Camera Sets
4   Canon Focus and Zoom control sets
4   Intercommunication Headphones
4   Special tripods

1 charriot plane and curve

1 Teleprompter with 2 reading systems (monitors

for individual speakers) 1 Lighting control table with memory – ARRI
Different microphones with diverse directional characteristics Audio components
Video components
General Intercommunication system

Video Régie

1 Recorder/Player BETACAM equipment
1 video digital mixer with 12 tracks
1 DVE equipment of digital effects for 3 channels
1 TBC Frame Synchroniser
2 Oscilloscope and Vectorscopy for control of cameras
2 Synchroniser Generators with Changeover
1 Intercommunication system with 4 places
1 Audio and Video Matrix with remote controls Video distributors

Audio Régie

1 audio mixer with 24 tracks
2 audio monitors
1 cd player
1 cassette deck
1   DAT
1   Mini-Disc
1   Level Detector for Audio Stereo Stereo Audio Distributors

Audio compressors Audio effects Microphones

Video Post-Production
3 Sony ES-7 on DVCAM/AVID hybrid non linear edition stations 1 Conventional BETACAM Edition Suite 2:1
Multi-track audio (8) on hard disc, Sound Scape type

3 ENG SETS:
Sony DSR-300 PK DVCAM Camcorder Digital Compact Report projectors
Various support materials (Batteries etc.)

3.10   Description of the Water Requirements for the project.

The requirements of water, separately for various matters are given in the following table.

Circulating   : Nil
Make-up   : Nil
Process   : Nil
Drinking   : 3,000 LPD

3.11 Description of the Steam requirement for the project.

A. Steam requirements and steam balance   : N.A.

B. Capacity and type of boiler with detailed specifications   : N.A.

C. Steam and energy diagram : N.A.

D.   Total energy generated / purchased (converted into M. K. : N.A. Cal) theoretical requirement of energy (in M. K. Cal)
at the various consumption stations and expected actual requirement at these stations.

E. If alternate processed are available, comparative energy : N.A.
consumption figures for the various processes. If the
project is energy intensive, possibility of choosing alternate process in order to make the project less energy
intensive.

F. Steps proposed to be taken by the company to improve   : N.A.
energy losses efficiency and reduce energy losses (such
as power factor improvement, power load manage-

ment,
optimising, illumination waste heat utilisation, etc.)

G. Scope for usage of solar / other renewable sources of   : N.A.
energy.

H. Any other measures contemplated in the direction of : N.A.
energy conservation and management.

3.123 Some information on Compressed air, fuel, etc.

Compressed Air

(a) Requirement   : N.A.
(b) Sources   : N.A.
(c) Arrangements proposed   : N.A.
(d) Cost at site with detailed calculations : N.A.

3.13   Details of the nature of atmospheric, soil and water pollution likely to be created by the project and the measures proposed for control of pollution. Indicate whether necessary permissions for the disposal of effluent have been obtained.

There shall not be any atmospheric pollution likely to be created by the project as there is no machine which has combustion resulting into air pollution or any chemical process which may release any gases which may result into an air pollution.

However, the use of DG set shall cause a small amount of air pollution. Considering the size of DG set, the pollution is within the permissible limits.

CAPE VERDE TELEVISION NETWORK

THE INDUSTRY

CHAPTER - 4

THE INDUSTRY

4.01 Description of the Television Industry in general.

The television industry can be described in the following broad categories. 1. Introduction
2.   Cable Networks

3.   DTH

4.   TV-media

5.   Earnings drivers

6.   Outlook

The detailed discussion in each of the categories is given below.

Introduction

The growing popularity of TV as a communication

medium has resulted in the TV media sector undergoing a rapid transformation. From the black and white days of state controlled TV Station, to the highly colourful tunes of Channel V and MTV, the medium has certainly undergone a phenomenal change. Given its popularity, percentage ad spend has also increased proportionately on this medium.

Media pie (%)

	1995	1997
TV	62.5	68.8
Radio	20.9	15
Press	16.6	16.2

Source NRS

Entry of new channels post 1991

All over the world the telecasting has witnessed entry of new channels to cater to the various needs of world audiences. Channels have been launched in English as well as other regional languages. In many countries of the world till 1991, the state owned TV Station ruled the roost, as other players were not allowed to uplink and broadcast. However channels such as CNN, Star TV and BBC, which were offshore companies, could circumvent these regulations and telecast their programs into any country of the world. Cable operators then relayed the same and made it available to the common man through the cable television network.

Like many other countries, the State machinery controlled television. It was used as a propaganda tool for the party in power, with the opposition always at the receiving end. The customer had very little choice. The first steps towards more user choice began during the 1980s, which had to be telecast to a wider audience. TV Station used satellite channels for the telecast and the T.V. network was launched as an international channel.

The sports telecast by Channel 9 in 1985 and the Gulf War in the late eighties all played small but important cameos in educating the international viewer. With liberalisation in 1992 and crumbling tariff barriers televisions (read as colour TVs) became more easily available. The media revolution had started.

Major satellite channels avidly watched by viewers are Star TV, Sony TV, Home TV, BBC & CNN. There are other regional language channels which are major players in their respective territories.

Most of the channels that could not attain popularity rapidly suffered, since their advertisement earnings were not sustainable. The first round of the media wars is over. Management changes, i.e. original promoters selling out to new management with deeper pockets, has become the order of the day. Alliances like the famous ESPN Star Sports arrangement also made headlines. Given the global trends of mergers and acquisitions, further consolidation is likely. Alliances and mergers make sense when the partners complement each other, like BBC and Discovery launched Animal Planet, CNBC and ABNI came together to launch a business channel called CNBC Asia.

Cable Networks

Antennas set up by either the end user or the cable operator receives the signals transmitted by the satellite. Local cable operators lay their own cables, set up control rooms, which can telecast 40 or more channels over a limited area. They charge the household a one- time connection charge of about US$.10 per point and a recurring monthly charge ranging between US$.1 to US$.5.

Initially, this was done in a very unorganised manner. The business required local knowledge and contacts, so every locality had its own cable operator. Collection was critical for the cable operator. For the end user, quality of telecast and a complete lack of standards became an issue. This lead to a shakeout and the formation of cable companies with money power which in turn tied-up with the local and small cable operators. Cable companies charge about US$1 per month to the local cable operators and support them with training and other infrastructure inputs. The business is immensely capital intensive and takes a long time to break even.

In many countries the operations of cable operators are regulated under the Cable TV Act which ensures that pornographic materials and other materials which are against culture and values or are detrimental to national interests do not get telecast. Recently this act has been amended to include foreign channels also.

Direct To Home

DTH is a new technology that circumvents the cable operators by directly delivering a bundle of channels to the end user. DTH involves transmission of encoded audio/ video signals (Ku band) via satellite. The end user needs an antenna to receive the signals and a decoder (set top box) to unscramble the encrypted signals. DTH services elsewhere in the world are Echostar and DirecTV (USA) and BskyB (Europe). Rupert Murdoch of Star TV fame owns BskyB.

The size of the antenna in DTH will be 1.5-2 ft in diameter, making it easy to install and transport. In conventional cable, since signals are in C band, an 8ft- diameter antenna is needed. The basic difference in the business model is the hardware costs in DTH. In a cable system, the user pays a one time connect fee and monthly rentals, while in DTH he has to invest in hardware.

The antenna will cost about US$200-300 and decoder will cost about US$200. The African viewer might be reluctant to incur such heavy installation costs. Quality of telecast in DTH is superior to Cable TV and viewer can receive up to 200 channels.

DTH will result in restructuring of the cable television industry. It will become imperative to have cash reserves to withstand the technology threat. Up gradation to fibre optic backbone will become necessary. A fibre optic network will cost about US$0.5mn per km as compared to US$0.1mn per km for coaxial cable. The stage is now ripe for consolidation.

TV Media features

In the Broadcasting business, it is only the industry leader who makes sizeable profits. The business is a game of asymmetrical payoffs. For instance, the top 5 channels account for 90% of ad spend.

Urbanisation and TV penetration is related. This may be due to the popularity of cable television that has resulted in increased colour TV sales. Rural penetration is low, although growing at a fast pace, because of dearth of specific program content to cater to that segment.

Liberalisation has resulted in the world viewer becoming more aware and conscious. This has resulted in the customer having more choice with the entry of a number of companies in different segments. Competition has resulted in companies increasing their marketing spend significantly.

Popularity of TV media is becoming higher. Increasing TV penetration leads to a reallocation of advertisement budgets with higher allocation for television at the cost of other medium.

TV channel operators use different business models to generate revenues. The critical component of any channel is the quality and type of programs they telecast. This determines their popularity, which in turn determines amount of advertisement revenues they can generate. They can do any one of the following:

Buy programming rights of program software from outside and collect advertisement revenue on their own. This model is followed by several TV companies, wherein they have a separate company in their fold, which develops all the content. The advantage is that re runs of serials/ programs become very profitable.

Selling time space to the producers for a fixed charge. Producers in turn are free to book advertisements at their own rates (there is an understanding on the time allocated for advertisement) and collect revenue. This is the basic model for many TV companies in which they sell prime time slots. The rights continue to be vested with the producer.

Earnings drives

The key factors that drive sector revenues are

Television penetration: Since the medium is television, increased television penetration will imply higher viewer ship. This will translate into higher advertisement spend allocation. This will also imply higher software production and demand for new programs.

Competition from other satellite channels would have an adverse impact on advertisement revenues, as advertisers have more choice in allocating ad budgets.

Government policies can have a big impact on the fortunes of the entire industry. When the DTH bill is passed in any country then, it will trigger a restructuring of the cable business.

Launching new channels targeted at specific segments, like regional channels within any country other areas having large pockets of ethnic population would lead to revenue growth. This will entail significant initial outlays.

Depreciation of the local currency would increase revenues as most of the program/ software companies export the programs overseas and payments are dollar denominated.

Advertisement revenue

As mentioned earlier, this is the primary source of income for TV channel operators. This revenue is directly co-related with the reach and viewer ship of a channel. Any channel's popularity depends on good quality programs, which is the software content. The business requires enormous initial investment in programs and revenues follow only with a time lag after the channel receives a minimum viewer acceptance.

Outlook

The sector has latent potential for growth on back of the exponential growth of cable TVs during the last 5 years. Television penetration in Africa is extremely low as compared to other developing countries like Malaysia, Pakistan, etc. in Asia. The number of channels has increased, implying higher demand for software programs.

Advertisement revenues, which are the barometer of channel popularity, will get dispersed over several competing channels. A shakeout is likely in both the channel and cable TV sectors. The biggest beneficiaries will be the content providers or the software houses. They will control the intellectual rights to the key element driving any channel's popularity.

Direct -to-Home, Digital Terrestrial Transmission and Conditional Access Cable Delivery have emerged as new delivery mechanisms. Breakthrough in technology would help open up avenues for these channels.

CAPE VERDE TELEVISION NETWORK

MARKETING

CHAPTER - 7

MARKETING

7.1   Description of the commercial viability of the project with regards to revenue generations.

Estimations shows that, after about 5 years of regular working, the CPLP Cultural Television Channel will be able to start exporting a great quantity of services and videos to Africa, Europe and South America. (See financial projections)

The economic feasibility of the project will occur with the commercialisation of several products, like:

1.   The sale and production of cultural programmes to various institutions like:
-     BBC
-     TV Culture Brazil
-     RTP
-     Discovery
Etc.

All over the world commercial networks for television cultural programmes have been created. Several television channels in different countries have regularly acquired programmes by producers spread out all over the world. Not only, several international organisations, like the World Bank and the United Nations, including FAO and Unesco, among many others, need programmes for public education, like programmes oriented to alert and to educate people concerning endemic and epidemic diseases. AIDS, malaria, tuberculosis are a very few examples.

These world institutions have made a great effort to develop television programmes with humanitarian objectives. But, in Africa – undoubtedly the continent most needed of such programmes – there is no television station, in present times, with capability to make face to such a need.

Therefore, the television teams that had been responsible for such programmes are, practically in its totality, placed in countries of the called First

World, strange to the local population's concrete reality.

Programmes focusing new agricultural techniques or even oriented to agricultural, commercial and industrial education are essential elements of such a repressed demand.

The price for each institutional campaign, with an average of 10 films produced per campaign, is of about USD.15.000$ and the capacity of the CPLP Cultural Channel will be as follows:

First Group* - from 2 to 4 campaigns per month.
Second Group – from 1 campaign in two months to 1 campaign per
month

This represents:
First Group   a potential annual income of about US$.540.000
Second Group a potential annual income of about US$.90.000

*First Group: Angola, Cape Verde, and Mozambique. Second Group: Guinea-Bissau, Sao Tome e Principe, East Timor (Lorosae). Reference Group: Brazil, Portugal.

Obs. In the Second Group, for the first six years of operation it is estimated a difference of potential annual income between the Angola unit and Cape Verde's and Mozambique's respective units, from US$.540.000 to about US$.378.000.

2. The sale and production of no cultural programmes to other television channels, like:
- series (novels)
- talk-shows Etc.

Many countries of the region need to produce television programmes, but do not have capability to do it. Thus, they seem themselves obliged to search expensive productions in Europe. With only one programme sold-three daily hours-the channels of the First Group* would receive in incomes the equivalent of about two times of the whole investment. The Second Group countries will not have conditions at the beginning to product cultural programmes to other television channels.

*First Group: Angola, Cape Verde, and Mozambique. Second Grope: Guinea-Bissau, Sao Tome e Principe, East Timor (Lorosae). Reference Group: Brazil, Portugal.

The price for this type of television programme is:

Auditory programs   US$.12.500 per programme
Talk-shows   US$.4.000 per programme

Interviews   US$.4.000 per programme

The capacity of production, of these programmes by the countries of the First Group will be:

In the FIRST PHASE:
Talk-shows   2 per month
Interviews   4 per month
In the FINAL PHASE:
Auditory programmes 6 per month
Talk-shows   20 per month
Interviews   20 per month

Thus, the potential income will be:

FIRST PHASE
Talk-shows   US$.96.000 per year
Interviews   US$.192.000 per year

FINAL PHASE
Auditory programmes US$.900.000 per year*
Talk-shows   US$.960.000 per year*
Interviews   US$.960.000 per year*

* the same difference as showed above.

The price of the novels is much higher. For each complete novel, with about 120 chapters, the price is about US$.2.500.000 and the capacity of the First Group of the CPLP Cultural Channel (in the final phase) will be of one novel per year.

3. There is the possibility of commercial use of the transmission time beyond the six hours reserved to culture and education.

The conventional day in television is of 18 hours, of which only 6 hours would be no commercial. We would have, therefore, 12 hours of no cultural transmission, which should be freely commercialised.
Even the period of six hours- divided into two sections of three hours, the first one dedicated to culture and second section of three hours to education-has a great potential for sponsoring.

Each commercial hour of transmission can include up to 12 minutes of advertising. The price for each 30 seconds of advertising is:

Noble time   from 6PM to 11PM   a b o u t US$.400 each 30"
Normal time   rest   about US$.200 each 30"

Concentrating the commercial programming in the noble time. With 4 hours of transmission in this period, we would have:

Educational television: from 7AM to 9AM
From 3PM to 4PM

Cultural television: from 4PM to 7PM
Commercial television: from 9AM to 3PM
From 7PM to 1AM

Thus, the commercial period would comprehend 6 hours in normal time, 4 hours in noble time and 2 hours in normal time again.

Begin 12 minutes per hour for advertising, the commercial period could have:
Noble time   48 minutes   24   films   US$.9.600
per day
Normal time   96 minutes   48   films   US$.9.600
per day
Total of the potential income........... US$.19.200
per day

> or US$.576.000 per month
> or US$.6.912.000 per year

It is believed that in the first phase of the project (after one year), the CPLP Cultural Television Channels of the First Group will be able to start with an income from commercial advertisement of about US$.1.000.000 per year-value which is predicted to increase in the follow months.

4. The rent of the studios to television teams of other countries can be another source of incomes. Many producers who cover events in Africa need to move many times to their countries of origin dur-

ing the video works, because there is not technical support in Africa. The same phenomenon happens with the cinematography and the journalism productions. It is not difficult to imagine, for example, the serious problems journalism teams have had, for example, with essential components like batteries, lighting, electronic components etc., which only can be easily find in Europe.

The period of renting of a television studio is of 12 hours. Each period has price of about US.4.000. After the first year, the studio should be rented for 12 hours per each period of 3 days. So, the rent per month, in this period (First Phase), will be able to generate incomes of about US$.120.000 per month in the First Group.

5. Support services to other television networks in all areas, including novels, mini series, docdramas series etc.

Many countries, principally in Africa, do not have technical conditions to develop this kind of programmes, but they have a strong internal repressed demand in this sense.

Each team, abroad, has a price of US$.1.000 per day (two people), after the costs of dislocation, hotel and meals. In the first months (first phase) it will be organised one team for works in other countries. In the final phase it is predicted to have up to 5 teams with such function. The capacity of each team is of about 20 days per month. Therefore, the support to other television networks-when took in its full potentiality-will be able to represent up to 20 days per month in the first phase. This represents a potential income of about US$.240.000 per year, and 100 days per month of works in the final phase, signifying about US$.1.200.000 per year of incomes, always referring to the First Group countries.

6. Colloquies, seminars, videoconferences and meetings of different natures.

The building of the CPLP Cultural Television Network will have a medium size auditorium, (First Group), with all conditions to receive seminars, colloquies and meetings of the most diverse nature. Such seminars, videoconferences etc. are important not only for the increase of the incomes, but also to attract specialists of the most different areas, being an important element for the development of all region as well as for the diffusion of the local, regional and continental cultural values.

The meetings, seminars and colloquies attract, in average, about 300 people per event. The price-excepting meals, hotels and transportation-per each participant is of about US$.50 per day. It will be a capacity for up to four events of this type per month, what could represent an income of about US$.60.000 per month or US$.720.000 per year (final phase-after 5 to 7 years).

The project should also turn possible.

-       classes of professional formation in diverse disciplines

-       support for a multimedia high technology centre (First Group)

-       support for a multimedia high cultural centre (Final Group)

CAPE VERDE TELEVISION NETWORK

PROFITABILITY & CASH FLOW

CHAPTER - 8

PROFITABILITY AND CASH FLOW

8.1   Estimation of cost of production and working results for the first five years of operation.

The estimated cost of production of working results for the first five years of operation are given in the chapter "Financial Projection".

8.2   Cash flow statement for the company as a whole, for five operating years of the project based on the estimates of working results.

A detailed cash flow statement for the company as a whole for five operating years is given in the chapter of "Financial Projection".

8.3   Projected balance sheet for five operating years for the company as a whole.

The balance sheet for five operation years for the company as a whole is given in the chapter of "Financial Projection".

CAPE VERDE TELEVISION NETWORK

ASSUMPTIONS

CHAPTER - 9

ASSUMPTIONS

1.   The Cape Verde Television Network shall generate the income from the following sources.

a)   Sale of cultural programmes.
b)   Sale of commercial programmes.
c)   Sale of advertisement time during transmission of TV programmes
d)   Hiring of studios.
e)   Supply of technical services
f)   Hiring of conference hall

The detailed assumptions for each of the above mentioned activities are as follows:

a)   Sale of cultural programs:

I.   It is assumed that four cultural programs shall be produced per month in the first year, which can be sold to other countries. This
figure will increase to 5 programs per month in the second year, six programs per month in the third year and so on.
II.   It is assumed that the selling price shall be US$.105 per program.

b)   Sale of commercial programs:

I.   The commercial programs consist of talk

shows, interviews and auditory programs.

II.   It is assumed that two talk shows shall be produced per month in the first year, three talk shows shall be produced in the third year, four talk shows per month shall be produced in the third year and
so on.

III.   It is assumed that the selling price of one talk show program shall
be US$.2800.

IV.   It is assumed that four interviews shall be produced per month in the first year, five interviews shall be produced in the third year, six interviews per month shall be produced in the third year and so on.

V.   It is assumed that the selling price of one interviews program shall be US$.2800.

VI.   It is assumed that no auditory programs shall be produced in the first and the second year. Only when the people have two years of experience then in the third year one auditory program shall be produced per month. In the fourth year two programs shall be produced per month and so on.

VII.   It is assumed that the selling price of one auditory program shall be UD$.8750.

c) Sale of advertisement time during the transmission of T.V. programs:

I.   There shall be a total transmission of 8 hours

per day in the first year. It will go on increasing by two hours in the second and subsequent years.

II.   Out of which advertisement shall be available for programs with transmission period of four hours in the first year. The programs in which advertisements could be available shall increase by two hours per year in the second and subsequent years.

III.   In one hour T.V. transmission, 12 minutes of advertisement shall be allowed.

IV.   Out of total T.V. time 33 % time shall be considered as prime time and remaining 67% shall be considered as non prime time, in the first, second and third years. From the fourth year onwards the prime-time advertisements shall remain constant.

V.   The advertisement rates shall be US$ 560 for one minute of prime time advertisement and US$ 245 for one minute of non prime time advertisement.

d)   Hiring of studios:

I.   It is assumed that the studios shall be taken on hire for a shift of 12 hours, two times in a month in the first year. It will increase to three times in a month in the second year.

II.   The rent per day shall be US$2800.

e)   Supply of technical services:-

I.   It is assumed that one team, consisting of

two technically qualified people shall be available in the first year. In the second year two such teams shall be available. In the third year three such teams shall be available and so on.

II.   It is assumed that in the first year the team shall be hired for 10 days in a month. In the second year the teams shall be hired for 12

days in a month. In the third year the teams shall be hired for 14 days in a month and so on.

III.   The rate of hire shall be US$.700 per team per day.

f)   Hiring of conference hall :-

I.   It is assumed that the conference hall shall be taken on hire for two times in a month in the first year, three times in a month in the

second year and four times in a month in the third year and so on.

II.   The hire charges shall be US$.350 per conference.

2.   The unit is planning transmission of three hours of educational programs, three hours of cultural programs and two hours of commercial programs per day in the first year. Thus, there shall be a total transmission of eight hours of programs per day in the first year. The total transmission hours shall increase by two hours per day in the second and subsequent years. Thus, in the second year there shall be total transmission of 10 hours per

day and in the third year there shall be a total transmission of 12 hours per day. The cost of software production is assumed at US$.1750 per hour.

3.   The cost of miscellaneous consumable items is assumed at about US$.25200 in the first year. This will increase by 10% in the second and subsequent years.

4.   The cost of stores and spares consumed is assumed at US$.29400 in the first year. It will increase by 10% in the second and subsequent years.

5.   The cost of repairs and maintenance is assumed at 0.1% of the total cost of plant and machinery in the first year. This cost will increase up to 0,3% in the fifth year.

6.   It is assumed that the unit shall use 1.000 units of power every day or 365.000 units of power in the first year. The consumption of power shall increase by 10% in the second and subsequent years. The cost of power is assumed at 3.5 cents for one unit.

7.   It is assumed that the unit shall consume 700 litters of diesel per month. The cost is assumed at 33 cents for one litter. It will increase by 10% in second and subsequent years.

8.   The unit shall have a technical staff of 35 people and administrative staff of 15 people.

9.   Depreciation is calculated on reducing balance method. The rate of depreciation is taken at 10% for building, 25% for plant and machinery, 10% for furniture and fixtures and 20% on vehicles.

10.   Advertisement expenditure is assumed at 0.2% of the sales.

11.   It is assumed that the stock of raw material and stores shall be 30 days of consumption, receivables shall be 45 days sales and creditors shall be 15 days purchases.

12.   The term loan shall be for a total period of 25 years. It shall have a moratorium of five years. It shall be repayable in 20 equal yearly instalments.

TELEVISÃO SONHO
Rede Planetária de Televisão Cultural para a Paz Mundial

PROJECT REPORT

OF

GUINEA BISSAU CULTURAL TELEVISION NETWORK

AT

GUINEA BISSAU

AFRICA

PROMOTERS

REPUBLIC OF GUINEA BISSAU

AFRICA

GUINEA BISSAU TELEVISION NETWORK

INTRODUCTION

CHAPTER - 1

INTRODUCTION

1.1   Description of the organisation through which this project is being taken up.

The CPLP is an international organisation, created in 17 of July of 1996, with Headquarters in Lisbon and consisting of the following State Members: Angola, Brazil, Cape-Verde, Guinea-Bissau, Mozambique, Portugal, Sao Tome e Principe and East Timor (Timor Lorosae).

The CPLP has as its objective the politic-diplomatic relationships between its Members, mainly referring to the United Nations and the World Bank, the relationship EU/MERCOSUL and the accomplishment of the Europe-Africa Summit.

The CPLP is also dedicated to the co-operation, particularly in the economic, social, cultural, legal and technical-scientific matters; as well as to the projects of promotion and broadcasting of the Portuguese Language, nominated to the improvement of the International Institute of the Portuguese Language and the creation of a Bibliographical Fund.

All its decisions are taken by consensus.

2.2   Description of the main reasons for setting up an African Cultural Television Network.

Africa is a continent submerged into dramatic problems. All these humanitarian problems have in the education and culture their most direct and objective root.

The African countries speaking Portuguese are among the African countries with the lowest level of development.

The whole situation is so dramatic that the implementation of the CPLP Cultural Television Network is more than urgent.

It is hoped that with this Television Network the people shall have better understanding of various diseases and other problems of life. e.g. AIDS is prevalent disease in Africa. But as the people are illiterate the information about the disease cannot be spread by any other method. In such cases the Television Network becomes a very effective audio-visual means of communication.

The project also intends to be a base of a massive

production of health campaigns, specially to reinforce the combat against the AIDS, the infect-contagious diseases- endemic or epidemic-through a process of global education of the local and regional populations.

The CPLP Cultural Television Network is a project for peace and its objective is not to compete with local television channels already installed, but to interchange with them, promoting a future global web dedicated to knowledge and development.

1.03 Description of the main objects of the project.

The major objective of the project is to assist the CPLP countries to develop the first African television network channel oriented to cultural and educational themes with the follow orientation:

-       diffusion of the Portuguese language
-       education in all areas
-       reinforcement of the local and regional cultures
-       formation of technical personal on communication and tele-education

-       formation and information for a better participation of the CPLP nations in the process of global development
-       to support the implantation of the Institute

of Portuguese Language

The development of such a project will create an irradiation pole from the countries with the project implementation to the whole African continent, constituting an African network of research and culture, discovering and valorising different cultural aspects and, consequently, reinforcing identity factors.

With the stronger cultural identity, it is expected to have lower levels of violence in social behaviour as well as a relevant increase of the educational performance in all its sectors.

Systematic educational objectives are complementary to the general orientation of this project, which is cultural in its fundament.
A very large positive impact is also to be expected with respect to several sectors of the economy, greatly improving the overall quality of services.

1.4   Description of the mission status of the project.

The mission of the project is:

-       to rescue lost values and cultural information

- to reinforce historical elements
- to rescue old artworks
- to give a new value to new artworks
- to communicate practical and objective information on health
- to communicate practical and objective information on agriculture
- to communicate practical and objective information on economy

1.05 Description of the philosophy of the project.

The CPLP Cultural Television Network will have in its philosophy, as a support to its mission, the follow fundaments:

- to work with local people
- be focused on quality
- intercommunication

Three words can define the CPLP Cultural Television Network philosophy:

People, quality and intercommunication

Following these three key words of its philosophy, a single sentence also defines its mission: "A Continent of Culture".

Two main sectors characterise its global structure:
1. Cultural Sector

1.1. To develop programmes on:
- Music
- Architecture (also vernacular)
- Archaeology
- Literature and poetry
- History
- Tourism
- Dance
- Plastic arts
- Astronomy
- Science in general
- gastronomy
- health
- anthropology
- local cultures
- others

1.2   The programmes can be of all types, the most important is to be attentive to communication. All programmes must reach the highest possible number of spectators.

2. Educational sector.
2.1   Systematic education in short practical courses.
2.2   Educational series with practical and objective information on:
- health care
- family
- agriculture
- managing

-     services
-     civil construction

As the CPLP Cultural Television Network must not be supported by the respective governments in the future, it will have a daily time of free commercial programmes up to 12 hours. This is a good solution to pay its costs and support its development.

1.06 Description the magnitude of the project.

The project proposes a Television transmission time of eight hours per day in the first year. It will increase by two hours per day in the second and subsequent years.

The project also envisages training of local people in all the related areas of Television. It is proposed to develop a local pool of talent which can be utilised not only in this particular country but also in other countries of Africa.

The total cost of the project of the Guinea Bissau Cultural Television Network is US$.10.600.000. Out of this it is proposed to obtain a soft loan of US$.8.800.000. The remaining amount of US$.1.800.000 shall be contributed by the Government of Guinea Bissau in terrain for the installation of the respective building, infrastructures and logistic support.

For the purpose of the Guinea Bissau Cultural Television Network a company shall be formed with the participation of the Government. The Contribution of the Government in the form of share capital shall be US$.1.800.000.

The term loan shall be on soft terms. The total period of the loan shall be 23 years. Out of which five years shall be a period of moratorium. The entire loan shall be repaid in the remaining 17 years, in 17 equal yearly instalments.

The Government of Guinea Bissau will give the sovereign guarantee for the referred loan.

GUINEA BISSAU TELEVISION NETWORK

THE COUNTRY

CHAPTER - 2

THE COUNTRY

2.1 - Description the level of development of Guinea Bissau.

One of the 20 poorest countries in the world, Guinea-Bissau depends mainly on farming and fishing.

Cashew crops have increased remarkably in the recent years, and the country now ranks sixth in cashew production.

Guinea- Bissau exports fish and seafood along with small amounts of peanuts, palm kernels, and timber. Rice is the major crop and staple food.

Trade reform and price liberalisation are the most successful part of the country's structural adjustment program under IMF sponsorship. The tightening of monetary policy and the development of the private sector have begun to reinvigorate the economy. Inflation dropped sharply in the first quarter of 1997. Membership in the WAMU (West African Monetary Union), begun in May 1997, should help support 5% annual growth and contribute to fiscal discipline. Because of high costs, the development of petroleum, phosphate, and other mineral resources is not a near-term prospect.

OFFICIAL NAME: Republic of Guinea-Bissau
CAPITAL: Bissau
SYSTEM OF GOVERNMENT: Multiparty Republic
AREA: 36,125 Sq Km (13,948 Sq Mi)
ESTIMATED 2000 POPULATION: 1,179,800

LOCATION & GEOGRAPHY: Guinea-Bissau is located on the coast of West Africa. It is bound by Guinea to the south and east, Senegal to the north and the Atlantic Ocean to the south- west. The territory consists of the mainland, the Bissagos archipelago that is a group of over 18 islands and various other coastal islands. The mainland terrain has a forested coastal plain typified by mangrove lined estuaries as well as a transitional savannah covered plateau which forms the Planalto de Bafata and Planalto de Gabu. The country is drained by a number of rivers that flow into the Atlantic Ocean and the principal rivers are the Cacheu or Farim, Mansoa, Geba, Corubal, Rio Grande and Cacine. Major Cities (pop. est.); Bissau 125,000, Bafata 14,000, Gabu 8,000 (1988). Land Use; forested 38%, pastures 38%, agricultural-cultivated 12%, other 12% (1993).

CLIMATE: Guinea-Bissau has a tropical climate with two seasons, a wet season from June to November with August the wettest month and a dry season from December to May with April and May the hottest months. The prevailing wind is the hot dust laden Harmattan which blows from the Sahara Desert in the west. Average annual precipitation in Bissau is 1,950 mm (77 inches) while average temperature ranges are from 24 degrees Celsius (75 degrees Fahrenheit) to 27 degrees Celsius (81 degrees Fahrenheit) all year.

PEOPLE: The majority of the population is of Black African origin and includes the following ethnic groups, the Balante who account for 27% while the Fulani account for 23%, the Malinke for 12%, the Mandyako for 11% and the Pepel for 10%. The Mestizos, Mulattoes and Assimilados who are of mixed racial descent, constitute the most important ethnic minorities representing around 2% of the population.

DEMOGRAPHIC/VITAL STATISTICS: Density; 27 persons per sq km (71 persons per sq mi) (1991). Urban-Rural; 19.9% urban, 80.1% rural (1990). Sex Distribution; 49.2% male, 50.8% female (1990). Life Expectancy at Birth; 39.9 years male, 43.1 years female (1990). Age Breakdown; 41% under 15, 25% 15 to 29, 17% 30 to 44, 11% 45 to 59, 5% 60 to 74, 1% 75 and over (1990). Birth Rate; 42.9 per 1,000 (1990). Death Rate; 23.0 per 1,000 (1990) . Increase

Rate; 19.9 per 1,000 (1990). Infant Mortality Rate; 151.0 per 1,000 live births (1990).

RELIGIONS: Over 65% of the population follow local native tribal beliefs while 30% are Muslims and the remainder are Christians, predominantly Roman Catholic.

LANGUAGES: The official language is Portuguese, although a Creole Patois is the trade and national language. The indigenous people all speak dialects derived from the Niger-Congo family of languages.

EDUCATION: Aged 25 or over and having attained: N/A. Literacy; literate population aged 15 or over 211,200 or 36.5% (1990).

CURRENCY: The official currency is the Peso (PG) divided into 100 Centavos.

ECONOMY: Gross National Product; USD $241,700,000 (1993). Public Debt; USD $633,600,000 (1993). Imports; USD $90,000,000 (1991). Exports; USD $23,000,000 (1991). Tourism Receipts; N/A. Balance of Trade; PG -293,146,000,000 (1994). Economically Active Population; 461,000 or 45.8% of total population (1992). Unemployed; N/A.

MAIN TRADING PARTNERS: Its main trading partners are Portugal, Egypt, Senegal, France, Spain,

Cape Verde, Germany and Algeria.

MAIN PRIMARY PRODUCTS: Bauxite, Cereals, Coconuts, Cotton, Fish, Palm Kernels, Peanuts, Rice, Timber.

MAJOR INDUSTRIES: Agriculture, Beverages, Construction, Fishing, Food

Processing, Forestry.

MAIN EXPORTS: Fish, Ground Nuts, Palm Kernels, Timber.

TRANSPORT: Railroads; nil. Roads; length 3,500 km (2,175 mi) (1989). Vehicles; cars 3,200 (1989), trucks and buses 2,400 (1989). Merchant Marine; vessels 18 (1990), dead weight tonnage 1,846 (1990). Air Transport; passenger-km 9,000,000 (5,592,000 passenger-mi) (1985), cargo ton-km 1,000,000 (685,000 short ton-mi) (1985).

COMMUNICATIONS: Daily Newspapers; total of 1 with a total circulation of 6,000 (1992). Radio; receivers 40,000 (1994). Television; N/A. Telephones; units 8,600 (1993).

MILITARY: 7,250 (1995) total active duty personnel with 93.8% army, 4.8% navy and 1.4% air force while military expenditure accounts for 3.3% (1992) of the Gross National Product (GNP).

GUINEA BISSAU TELEVISION NETWORK

THE PROJECT

CHAPTER - 3

THE PROJECT

3.1   Description of the initial steps to be taken for the implementation of the project.

The first objective of the CPLP Cultural Television Network is to install its structures in the following countries:

- Sao Tome e Principe
- Cape Verde
- Mozambique
- Guinea-Bissau
- Angola
- Timor

Thus, these six countries will structure the basic network of the whole project.

Angola, Mozambique and Cape Verde will form a FIRST GROUP; Sao Tome e Principe, Guinea-Bissau and East Timor will form a SECOND GROUP; and, finally, Brazil and Portugal will form a called REFERENCE GROUP.

However, the project will start at the same time in all countries. A more precise description of the meaning of each group is made below.

The basic physical components of the project are:

1.   The creation of companies, in each of the States of the CPLP, oriented to the objectives above referred.

2.   The State of each country – who will be the responsible and guarantee of the payment of the financing – will also have a participation in the companies.

3.   All strategies will be oriented to lowest costs in long term.

4.   The whole project will be developed by phases.

5.   The timetable of transmission will be defined case by case.

6.   The project will include:

6.1.   a complete plan for an equipment network.

6.2.   a complete plan for a software network.

6.3.   all procurements

6.4.   equipment transportation and installation

6.5.   all configurations

6.6.   creation of the infographic system

6.7.   technical formation of the personal

6.8.   architecture and engineer projects for the building

6.9.   construction of the building organised in phases

1.2   Description of the procedure by which the necessary programmes will be produced for transmission on the channel.

The television programs are known as software in this field. The unit is planning transmission of three hours of educational programs, three hours of cultural programs and two hours of commercial programs per day in the first year. Thus, there shall be a total transmission of eight hours of programs per day in the first year. The total transmission hours shall increase by two hours per day in the second and subsequent years. Thus, in the second year there shall be total transmission of 10 hours per day and in the third year there shall be a total transmission of 12 hours per day. The cost of software production is assumed at US$.2.500 per hour. The amount of US$.2.500 per hour is taken based on cost of production of such programmes in Portugal.

To contract a similar system from other television channel, only for one country, for a transmission period of 6 hours daily, the cost in, market (eg. Portugal*) is of about US$.500.000 per month. This cost represents a total of about US$.6.000.000 per year. In 16 years, excluding the first year of implementation, the costs for contract the six daily hours of transmission from other television channel, in present market price, is of about US$.100.000.000. The six countries would have an estimated cost of about US$.600.000.000 in a 16 years period of time.

These costs are based on the prices presently used in Portugal, as for example the programme Hora Viva – Seguranca em Directo, transmitted every day, excluding weekends, from 7 to 10 AM.

Another example is the channel Canal Noticias Lisboa CNL. Only in its first year of installation, it was expended about US$.10.000.000, and the equipment as well as the building were not property of the channel, but rented.

In fact, however the project will be implemented by phases, after 5 to 10 years of regular work, it is predicted to have reached a similar value in incomes as the examples above.

*value took in the Portuguese market, below the average of the European prices in about 30%.

The first ten months of works will be oriented to:
-        design of the equipment and software structures

- start the first productions
- elaboration of the projects of architecture and engineering
- procurement and acquisition of all equipment
- construction and supervision of the works
- installation and configuration of the equipment

So, the first year must be used to the implantation of the project. However, as to accelerate the chronogram, some operations should be started before this timing, temporarily installed in a different building.

The first phase will be oriented to:
- video productions (external and internal)
- edition
- elaboration of cost programmes

3.3   Description of the requirements of Power, Fuel and Water. POWER
It is assumed that the unit shall use one thousand units of power of everyday. Thus, the unit shall use 365.000 units of power in a year. The costs of power is assumed at two cents for one unit. The consumption of power shall increase by 10% every year.

FUEL

The unit shall use D.G. set for generation of electricity, when there is an interruption in the supply. The hours of disruption in the normal supply can not be predicted accurately. Hence, the hours for which the D.G. sets shall be used also can not be predicted accurately. However, it is assumed that the unit shall utilise one thousand litters of diesel per month. The cost is assumed in about 13 cents for one litter. Hence, the total cost for one thousand litters shall be US$.130. The consumption of diesel shall increase by 10% every year.

WATER

The unit does not need water for fixed commercial operation. However, it shall have a strength of 50 people as employees. Water shall be required for drinking and sanitation purposes. It is assumed that 3000 litters of water shall be required everyday. The cost of obtaining this water is assumed at US$.4 everyday. The consumption of water shall increase by 10% every year.

3.4   Description of the manpower requirement of the project and description of the steps taken by the company to train the manpower.

The total staff strength shall be 17 people in the first year.

The system of modern TV Station is designed to achieve sustained operating efficiency and transmission. This, of course, entails a certain degree of sophistication in the production and transmission. However, considering the socio-economic situation in Guinea Bissau, a reasonable balance has to be struck to obtain optimum performance and at the same time create gainful employment. While working out the manpower requirement for this project to be kept on direct rolls of the company totalling 50 staff and operators, the above consideration have been kept in mind.

Manpower requirement

The direct manpower required for the proposed Unit is about US$.720.000 per year. The manpower requirement as indicated in this chapter has been planned keeping in view the following guide-lines:

Effective co-ordination among the various departments. Judicious distribution of responsibilities.

Capacity utilisation of the TV Station with optimum manpower.

Details of manpower requirement for the TV Station is given in the financial section.

In line with the prevailing practice, the Security guards, office peons and unskilled labour etc., are normally employed on contractual/casual basis, and their cost has to be included in the Factory over-heads. However in this report provision has been made to employ the above staff and their salaried have been included.

The various departments proposed shall be under the direct responsibility of the Station Director. He shall be assisted by Deputy Directors, who will look after the complete TV Station and its various day to day activities and Marketing. They shall be responsible for achieving the envisaged targets and sales forecasts. They shall be assisted by a team of Managers from production, marketing and accounts.

To run this project a labour force which shall usually be composed of unskilled and skilled workers shall be employed. The first are those who do not undergo any king of specific training or education, while the latter have to do so in order to master their jobs.

When evaluating an investment project from an employment point of view, its impact on both unskilled and skilled labour has been taken into account. Not only direct employment, but also indirect employment has been considered. Direct employment refers to the new employment opportunities created within the project; indirect employment concerns job opportunities created in other projects linked with the project which is being formulated.

The implementation of large and sophisticated projects generally contributes to the development of local skills and capabilities in a country. Furthermore, they help to change traditional values, attitudes and the behaviour of the society, to build up an enterprising spirit among the people, to develop a desire for changing and improving the existing conditions of life, to introduce better work discipline and thus to change the very pattern and basis of economic development. The TV industry is already well established in Africa. Location of TV industry activities in Africa, has certain favourable factors and advantages, in setting up this type of industry, as availability of acceptable level of education among the supervisory staff adequate technical and managerial skills developed over a long period, and availability of cheap labour are assured.

The organisation structure will vary from TV Station to TV Station in the industry and as such the pattern proposed herein can be considered only as suggestive and provisional.

A well-knit organisation structure headed by a Station Director and Manager, with the supporting staff will be developed progressively during project implementation. Soon after the plant becomes operative, a good number of project staff will be absorbed in the organisation. Certain additional staff also get added to ensure smooth and efficient management of the operating unit.

The Manager Production will have a degree in communication and will be accountable and answerable to the Station Director for all TV operations including planning, production, material management, TV utilities, quality control, production cost and budgetary control, TV safety, discipline and layout relations. His main function should be to ensure achievement for quality and quantity targets of production at reasonable cost and should constantly strive to improve TV performance. In the discharge of his multifarious duties and responsibilities he will be assisted by supervisors and adequate staff for day to day activities.

His duties are to ensure that targets are maintained through effective utilisation of 3 M's viz: Machinery, Men and Materials. These targets should be translated in terms of targets for individuals under him such as supervisors, skilled workers etc. He has to ensure that the equipment at his disposal gets prompt attention on breakdown is adhered to strictly. He has to further ensure that there is a strict quality control exercised over raw materials also. He will be assisted by Deputy Manager etc. who and will report to the General Manager.

GENERAL MANAGER (Comm.) He will report to the Station Director. His main duties will be prepare sales forecasts and budgets, study and monitor the export market for the company's products and advice on ways and means of increasing sales. He will also ensure that consumer complains are solved in an appropriate manner. He will also ensure that the right materials are available for production at the right time.

He will be responsible for all aspects connected with the export procedures and keep the management updated on all matters relating to the Govt. policies on polymers and Exports and above all the world market.

COMPANY SECRETARY & FINANCIAL CONTROLLER. He will report to the Station Director and will be responsible for the departments of administration, financial planning, budgetary control, cost accounting, tax management payroll accounting, etc. and all personnel matters. He will be assisted in their duties by their respective assistants to assist in day to day activities.

Manpower planning and production

The central issue here may be one of scale. Production is staffed bye personnel, but the process may be labour-intensive or capital-intensive. Either way, planning will include:

1.   Analysis of labour supply and demand factors in relation to skills and training needs.
2.   Procedure for manpower recruitment, together with selection processes.
3.   The formation of industrial relations policies necessary for effective work place bargaining, disciplinary measures and dismissal procedures.
4.   Analysis of the effective use of human resources.
5.   Conditions necessary to maintain adequate levels of motivation. The scale of the problem is likely to be directly proportional to the method of production.

The availability of main persons is not going to be easy since TV industry is currently in its infancy in Africa.

Training needs

The selection and training of the required manpower for the proposed project has to be planned in advance.

The key personnel should be selected and trained suitably. The training would be carried out in the following manner:

Basic training on the concept of TV industry before construction begins with visits to similar TV Stations in other countries. On site training during the construction phase of the project. On job training during the commissioning phase of the project. On job training during operation of the TV immediately after commissioning. The training of the key personnel such as Station Director should be carried out in all the phases. The training of other operating personnel should be suitably carried out during the construction, erection and operation phases in addition to training them by visits to similar plants operating in other countries.

Besides training the key operating staff described above, in TV training should also be given to other employees at skilled operating level to enable them to understand the process equipment in the project and prepare them to operate a maintain their respective sections safely, efficiently and skilfully. The above training should be carried out during construction, commissioning and operating phases of the project.

Training is necessary in order to enable personnel to acquire the skills and knowledge necessary to perform a task to an acceptable standard. The length of the training period and training methods will, of course, vary from job to job. Training is essentially a learning process, and in order that progress can be successfully monitored certain conditions are necessary.

1.   The training needs of both the individual and the organisation shall be identified and analysed.

2.   Targets and standards shall be set for the trainee, which are within his capabilities.

3.   The pace of the training programme should reflect the trainee's ability to maintain progress in properly absorbing the same.

4.   The trainee shall receive regular feedback of results. Any problem areas shall be highlighted, discussed and resolved.

5.   As the trainee progresses the amount of information provided shall be gradually reduced, thus inducing a feeling of independence and competence.

It is common place to find a wide variety of tasks in an organisation and each will require varying degrees of skill, effort and responsibility. This being so, it is inevitable that rates of pay will also vary and the differentials between the jobs will reflect their relative values. However, other factors such as local market conditions, bargaining strengths and traditions also influence a company's payment structure and a great deal of planning is required if rationalisation is to be achieved. One technique which has been successfully adopted by many companies to establish an equitable wage structure is job evaluation. In a job evaluation exercise a comparison is made of common criteria over a range of jobs, and the resulting analysis may be linked to a points allocation or job ranking system, and hence to a wage scale.

In conducting a job evaluation exercise it is important to cover a reasonable variety of tasks within the whole spectrum. For each, a job description is prepared setting out details of the duties and responsibilities undertaken by the employee together with a statement about his working conditions. Very often this task is undertaken by work study personnel since, they are responsible for determining methods of operation and evaluating the work content of the job. Each job is assessed factor by factor, resulting in a comprehensive comparative analysis.

The individual is the most important resources of any company and only people who are well trained, well motivated and adequately rewarded will provide a positive and synergistic contribution towards the company's objective and its organisational health.

In most cases, the factors, which may be weighted according to relative value, are as follows:

1.	Skill-education, experience and training.
2.	Effort-both physical and mental.
3.	Responsibility-for equipment, materials, initiative etc.
4.	Working conditions-general conditions, risk of accident and injury.

Pay policies affect not only individual employees but the whole organisation, and the rewards and objectives vary at different levels within the enterprise.

Industrial relations

An effective industrial relations policy is important, since is the system through which employees take part in decision-making and in many instances it affects the while atmosphere of employer/employee relationships. An industrial relations policy is essentially a set of rules whose determine procedures for negotiation on such matters as:

1.	Wage and salary scales.
2.	Terms and conditions of work.
3.	Disputes and grievances.
4.	Recruitment and dismissal.
5.	Other issues of mutual interest, e.g. closed shop, redundancies and joint consultation.

In order to promote an atmosphere of co-operation, and to minimise conflict, the needs of management and work people must be recognised by both sides. Trade unions exit to protect the interests of their members and improve their working conditions. Management, while aware of the pressures and constraints imposed by the trade unions, have a duty to maximise the use of resources at their disposal, which may be expressed in relation to profitability, return on investment, level of service, sales volume, market share and cost-effectiveness. The strategies adopted in attempting to solve industrial relations problems will vary from company to company, and indeed from union to union, but there is no doubt that they will be influenced by both internal and external factors.

Internal factors

1.   The attitudes of employees to management, and management to employees.
2.   The leadership style of management.
3.   The bargaining strength of both parties.
4.   The number of negotiating bodies.
5.   The prosperity of the company.

External factors

1.   The extent to which parent boards influence company management, and district officials

influence or control local shop stewards.

2. Whether or not bargaining is conducted at plant, local or national level.

3. Government policy towards industrial relations.

4. The economic situations nationally, locally or within the company itself.

3.5 Describe the project implementation schedule.

Implementation of this project is a challenging task and calls for meticulous planning, scheduling and monitoring to realise the project goals within the budgeted cost and time frame. The goal can be achieved by adopting modern project management techniques.

To implement this project adequately, a team of engineers and project personnel having requisite education and experience are being appointed, to whom a detailed Work Breakdown Structure (WBS) in a logical order of activities, shall be supplied shortly, keeping in view cost estimation, scheduling, and to help monitor and control of the project. It is proposed to be formulated in conjunction with the objectives of each activity and goal settings. Project shall be programmed and controlled by network analysis techniques. Before the application of the network analysis techniques, the project personnel shall be acquainted with their capabilities in saving time, resources and costs. The training of project personnel and engineers at levels shall be provided for proper control of project progress and taking of timely corrective actions to re-align these efforts to meet ore stated objectives. It is proposed to gear programming and control system, i.e. project implementation system, which will ensure an integrated approach to project implementation. Project management activities shall be determined in advance and all activities carried out be project personnel as well as those to be contracted shall be identified.

Responsibility for project implementation shall be clearly defined. The forms of project organisation range from project oriented to functional organisation, while most of the cases are combinations of the two, with certain adaptations to prevailing conditions. It is impossible to over emphasise the importance of establishing a team of a task force for implementing the project with a designated leader to co-ordinate and guide its functions.

Project manager: Shall be responsible to the Board of Directors. The project manager shall be responsible for guiding and co-ordinating the efforts of all parties engaged in implementing the project, obtaining necessary government approvals on contracts. He is to control the project organisation with the promoters as well as with other agencies and organisations interested in the project. The manager shall have some staff to assist him, especially in checking expenditures to date and determining the present and future cost overrun or under run so that the project manager can take or propose to the Board pertinent corrective measures.

The network shall cover the pre-construction phase of the project indicating major administrative processes, since experience shows that some of them have frequently involved lengthy delays. In other words, it shall include the aggregate activities to be carried out be the principal parties participating in the implementation process.

The project budget shall then be prepared. Part or periodic payments to contractors which might be made at the end of certain intervals (e.g. weekly or monthly) throughout the time horizon of project implementation, shall be made by summing up activity costs per unit of time, which may be a week & month, and computing the cumulative cost at the end of each time interval. For the activities that are in process and are contacted or sub-contracted, the assumption of a linear time cost activity relationship shall be used for the sake of simplicity. In other words, expenditures are uniformly distributed throughout the duration of the activity.

PROJECT SCHEDULE

After an investment decision is taken, the main machinery and long delivery items must be ordered out at the earliest, forming the first major step in implementation of the project. It is foreseen that an engineering consultant will be appointed for carrying out the detailed engineering including basic engineering and procurement assistance to the client. It is also assumed that reputed and experienced contractors with adequate resources viz., men, materials, tools and tackles etc. will be engaged for execution of the construction and erection work. The purchase packages for auxiliaries shall be kept minimum so as to reduce the co-ordination efforts to the minimum. A great deal of co-ordination is required for constructing/erecting the new units. This task is feasible, provided the major activities of the project are co-ordinated and completed in the duration specified to achieve the respective milestones in time.

STRATEGY FOR TIMELY EXECUTION

It is important to deploy a team of experienced personnel for project execution and select the external agencies with due care for rendering the services and supply of equipment for the project. The project activities must be identified, planned and scheduled, and the progress monitored for timely project implementation. All the inputs to the project including financial resources must be identified and their inflow planned and arranged in time.

The project must be managed professionally with necessary co-ordination among the various agencies and requisite decisions taken promptly.

Establishment of an effective monitoring procedure for progress review and co-ordination.

However this tentative schedule, many activities will begin in the 6th month after the start of the financing, using different locations and training some partial personal. This will turn possible the strength of the dynamics in the whole process as well as the shorting of the estimated timing.

In short, the following key factors would constitute the broad strategy for timely execution of all activities in a pre-determined manner as per schedule shown in the bar chart as to reach at a basis of regular production.

Early selection of an effective in-house technical team (TASK FORCE) by Government of Guinea Bissau, headed by a Project Manager for planning and executing the project.

i)   Proper choice of external agencies such as consultants for Project Engineering., Machinery suppliers, Construction Agencies etc. keeping in view their reputation/past performance and working experience in their fields.

ii)   Adequate use of computer-based PERT/ CPM techniques for project planning, scheduling and monitoring.

3.6   Description of the initial project Flowchart.

The project implementation phase embraces the period from the decision to start the project to the beginning of the commercial production. It includes a number of stages including negotiations and contracting, project design, construction and start-up.

3.7   Description of the requirement of Land.

For this project the Government of Guinea Bissau will participate with about 25.000 square meters of land. In selecting the land the following criteria should be followed.

1.   The land should be near to the main city of Bissau.

2.   The land should be properly connected by road.

3.   It should also have proximity to many end user clients.

4.   It should also meet various Government policy of

       achieving the social objectives.

2.08 Description of the requirement of Building.

The total requirement of building is as follows.

a. Adm. Building   400 Square Meters

b. Studios –2 (Two)   1000 Square Meters

c. Other Misc. Building   200 Square Meters

Total Square Meters: approximately 1600 Square Meters.

Cost for construction of about US$.750 per square meter (including air conditioning systems, primary electrical cabin etc.).

2.9   Description of some requirements for Plant & Machinery and their basis of selection.

The plant and machinery have been selected giving due consideration to the sophisticated nature of technology required. Detailed discussions were carried out with the foreign suppliers to ensure that the required capacities are practical with minimum capital and operating costs before the machinery were finally selected.

The machinery shall be selected from the most reputed foreign manufacturers of complete range of equipment. Keeping in view the following main factors:

a)   Past performance
b)   Existing machinery in Africa & abroad.
c)   Sales and service facilities in Africa

The main equipment is as follows.

STUDIO

1.   Post-production equipment – non-linear post-production, two ES7 stations.
2.   ENG – 2 DVCAM Camcorders
3.   Régie with 2 DXC-D30PK1 Digital Cameras - for cold light installation.

4.   Flexicart.
5.   Computer graphics suite.

Other technical data:

The whole installation is characterised by:

1. Informational web with UTP level, 5 cables.

2.   Phone system.

3.   Specific furniture.

Discrimination of Technical Data:

Studio
Cold lighting system.
2 Camera channels composed by:
1   DXC-D30PK Sony Digital Cameras 2 J18 Canon Zoom Lens

2 CA-537P Sony Camera Adapters
1   DXF-50 Sony Studio Viewfinder's
1   CCU-M3P Sony Control Camera Units 2 RM-7P Sony remote CCU controls

2 50M Sony Multicore Camera Sets
2 Canon Focus and Zoom control sets
2 Intercommunication Headphones
2 Special tripods
1   charriot plane and curve

1   Teleprompter with 2 reading systems (monitors for individual speakers) Different microphones with diverse directional characteristics

Audio components Video components

Video Régie

1 Recorder/Player BETACAM equipment
1 video digital mixer with 12 tracks
1 DVE equipment of digital effects for 3 channels
1 TBC Frame Synchroniser
1 Oscilloscope and Vectorscopy for control of cameras
1 Synchroniser Generators with Changeover
1 Intercommunication system with 4 places
1 Audio and Video Matrix with remote controls Video distributors

Audio Régie

1 audio mixer with 24 tracks
1 audio monitors
1 cd player
1 cassette deck
1   DAT
1   Mini-Disc
1   Level Detector for Audio Stereo Stereo Audio Distributors

Audio compressors Audio effects Microphones

Video Post-Production
2 Sony ES-7 on DVCAM/AVID hybrid non linear edition stations 1 Conventional BETACAM Edition Suite 2:1
Multi-track audio (8) on hard disc, Sound Scape type

3 ENG SETS:
Sony DSR-300 PK DVCAM Camcorder Digital Compact Report projectors
Various support materials (Batteries etc.)

3.10   Description of the Water Requirements for the project.

The requirements of water, separately for various matters are given in the following table.

Circulating	: Nil
Make-up	: Nil
Process	: Nil
Drinking	: 3,000 LPD

3.11 Description of the Steam requirement for the project.

A. Steam requirements and steam balance   : N.A.

B. Capacity and type of boiler with detailed specifications   : N.A.

C. Steam and energy diagram : N.A.

D.   Total energy generated / purchased (converted into M. K. : N.A. Cal) theoretical requirement of energy (in M. K. Cal)

at the various consumption stations and expected actual requirement at these stations.

E. If alternate processed are available, comparative energy : N.A.
consumption figures for the various processes. If the
project is energy intensive, possibility of choosing alternate process in order to make the project less energy
intensive.

F. Steps proposed to be taken by the company to improve : N.A.
energy losses efficiency and reduce energy losses (such
as power factor improvement, power load management,
optimising, illumination waste heat utilisation,

etc.)

G. Scope for usage of solar / other renewable sources of   : N.A.
energy.

H. Any other measures contemplated in the direction of : N.A.
energy conservation and management.

3.123 Some information on Compressed air, fuel, etc.

Compressed Air

(a) Requirement   : N.A.
(b) Sources   : N.A.
(c) Arrangements proposed   : N.A.
(d) Cost at site with detailed calculations : N.A.

3.13   Details of the nature of atmospheric, soil and water pollution likely to be created by the project and the measures proposed for control of pollution. Indicate whether necessary permissions for the disposal of effluent have been obtained.

There shall not be any atmospheric pollution likely to be created by the project as there is no machine which has combustion resulting into air pollution or any chemical process which may release any gases which may result into an air pollution.

However, the use of DG set shall cause a small amount of air pollution. Considering the size of DG set, the pollution is within the permissible limits.

GUINEA BISSAU TELEVISION NETWORK

THE INDUSTRY

CHAPTER - 4

THE INDUSTRY

4.01 Description of the Television Industry in general.

The television industry can be described in the following broad categories. 1. Introduction
2.   Cable Networks

3.   DTH

4.   TV-media

5.   Earnings drivers

6.   Outlook

The detailed discussion in each of the categories is given below.

Introduction

The growing popularity of TV as a communication medium has resulted in the TV media sector undergoing a rapid transformation. From the black and white days of state controlled TV Station, to the highly colourful tunes of Channel V and MTV, the medium has certainly undergone a phenomenal change. Given its popularity, percentage ad spend has also increased proportionately on this medium.

Media pie (%)

	1995	1997
TV	62.5	68.8
Radio	20.9	15
Press	16.6	16.2

Source NRS

Entry of new channels post 1991

All over the world the telecasting has witnessed entry of new channels to cater to the various needs of world audiences. Channels have been launched in English as well as other regional languages. In many countries of the world till 1991, the state owned TV Station ruled the roost, as other players were not allowed to uplink and broadcast. However channels such as CNN, Star TV and BBC, which were offshore companies, could circumvent these regulations and telecast their programs into any country of the world. Cable operators then relayed the same and made it available to the common man through the cable television network.

Like many other countries, the State machinery controlled television. It was used as a propaganda tool for the party in power, with the opposition always at the receiving end. The customer had very little choice. The first steps towards more user choice began during the 1980s, which had to be telecast to a wider audience. TV Station used satellite channels for the telecast and the T.V. network was launched as an international channel.

The sports telecast by Channel 9 in 1985 and the Gulf War in the late eighties all played small but important cameos in educating the international viewer. With liberalisation in 1992 and crumbling tariff barriers televisions (read as colour TVs) became more easily available. The media revolution had started.

Major satellite channels avidly watched by viewers are Star TV, Sony TV, Home TV, BBC & CNN. There are other regional language channels which are major players in their respective territories.

Most of the channels that could not attain popularity rapidly suffered, since their advertisement earnings were not sustainable. The first round of the media wars is over. Management changes, i.e. original promoters selling out to new management with deeper pockets, has become the order of the day. Alliances like the famous ESPN Star Sports arrangement also made headlines. Given the global trends of mergers and acquisitions, further consolidation is likely. Alliances and mergers make sense when the partners complement each other, like BBC and Discovery launched Animal Planet, CNBC and ABNI came together to launch a business channel called CNBC Asia.

Cable Networks

Antennas set up by either the end user or the cable operator receives the signals transmitted by the satellite. Local cable operators lay their own cables, set up control rooms, which can telecast 40 or more channels over a limited area. They charge the household a one- time connection charge of about US$.10 per point and a recurring monthly charge ranging between US$.1 to US$.5.

Initially, this was done in a very unorganised manner. The business required local knowledge and contacts, so every locality had its own cable operator. Collection was critical for the cable operator. For the end user, quality of telecast and a complete lack of standards became an issue. This lead to a shakeout and the formation of cable companies with money power which in turn tied-up with the local and small cable operators. Cable companies charge about US$1 per month to the local cable operators and support them with training and other infrastructure inputs. The business is immensely capital intensive and takes a long time to break even.

In many countries the operations of cable operators are regulated under the Cable TV Act which ensures that pornographic materials and other materials which are against culture and values or are detrimental to national interests do not get telecast. Recently this act has been amended to include foreign channels also.

Direct To Home

DTH is a new technology that circumvents the cable operators by directly delivering a bundle of channels to the end user. DTH involves transmission of encoded audio/ video signals (Ku band) via satellite. The end user needs an antenna to receive the signals and a decoder (set top box) to unscramble the encrypted signals. DTH services elsewhere in the world are Echostar and DirecTV (USA) and BskyB (Europe). Rupert Murdoch of Star TV fame owns BskyB.

The size of the antenna in DTH will be 1.5-2 ft in diameter, making it easy to install and transport. In conventional cable, since signals are in C band, an 8ft- diameter antenna is needed. The basic difference in the business model is the hardware costs in DTH. In a cable system, the user pays a one time connect fee and monthly rentals, while in DTH he has to invest in hardware.

The antenna will cost about US$200-300 and decoder will cost about US$200. The African viewer might be reluctant to incur such heavy installation costs. Quality of telecast in DTH is superior to Cable TV and viewer can receive up to 200 channels.

DTH will result in restructuring of the cable television industry. It will become imperative to have cash reserves to withstand the technology threat. Up gradation to fibre optic backbone will become necessary. A fibre optic network will cost about US$0.5mn per km as compared to US$0.1mn per km for coaxial cable. The stage is now ripe for consolidation.

TV Media features

In the Broadcasting business, it is only the industry leader who makes sizeable profits. The business is a game of asymmetrical payoffs. For instance, the top 5 channels account for 90% of ad spend.

Urbanisation and TV penetration is related. This may be due to the popularity of cable television that has resulted in increased colour TV sales. Rural penetration is low, although growing at a fast pace, because of dearth of specific program content to cater to that segment.

Liberalisation has resulted in the world viewer becoming more aware and conscious. This has resulted in the customer having more choice with the entry of a number of companies in different segments. Competition has resulted in companies increasing their marketing spend significantly.

Popularity of TV media is becoming higher. Increasing TV penetration leads to a reallocation of advertisement budgets with higher allocation for television at the cost of other medium.

TV channel operators use different business models to generate revenues. The critical component of any channel is the quality and type of programs they telecast. This determines their popularity, which in turn determines amount of advertisement revenues they can generate. They can do any one of the following:

Buy programming rights of program software from outside and collect advertisement revenue on their own. This model is followed by several TV companies, wherein they have a separate company in their fold, which develops all the content. The advantage is that re runs of serials/ programs become very profitable.

Selling time space to the producers for a fixed charge. Producers in turn are free to book advertisements at their own rates (there is an understanding on the time allocated for advertisement) and collect revenue. This is the basic model for many TV companies in which they sell prime time slots. The rights continue to be vested with the producer.

Earnings drives

The key factors that drive sector revenues are

Television penetration: Since the medium is television, increased television penetration will imply higher viewer ship. This will translate into higher advertisement spend allocation. This will also imply higher software production and demand for new programs.

Competition from other satellite channels would have an adverse impact on advertisement revenues, as advertisers have more choice in allocating ad budgets.

Government policies can have a big impact on the fortunes of the entire industry. When the DTH bill is passed in any country then, it will trigger a restructuring of the cable business.

Launching new channels targeted at specific segments, like regional channels within any country other areas having large pockets of ethnic population would lead to revenue growth. This will entail significant initial outlays.

Depreciation of the local currency would increase revenues as most of the program/ software companies export the programs overseas and payments are dollar denominated.

Advertisement revenue

As mentioned earlier, this is the primary source of income for TV channel operators. This revenue is directly co-related with the reach and viewer ship of a channel. Any channel's popularity depends on good quality programs, which is the software content. The business requires enormous initial investment in programs and revenues follow only with a time lag after the channel receives a minimum viewer acceptance.

Outlook

The sector has latent potential for growth on back of the exponential growth of cable TVs during the last 5 years. Television penetration in Africa is extremely low as compared to other developing countries like Malaysia, Pakistan, etc. in Asia. The number of channels has increased, implying higher demand for software programs.

Advertisement revenues, which are the barometer of channel popularity, will get dispersed over several competing channels. A shakeout is likely in both the channel and cable TV sectors. The biggest beneficiaries will be the content providers or the software houses. They will control the intellectual rights to the key element driving any channel's popularity.

Direct-to-Home, Digital Terrestrial Transmission and Conditional Access Cable Delivery have emerged as new delivery mechanisms. Breakthrough in technology would help open up avenues for these channels.

GUINEA BISSAU TELEVISION NETWORK

MARKETING

CHAPTER - 7

MARKETING

7.1 Description of the commercial viability of the project with regards to revenue generations.

Estimations shows that, after about 5 years of regular working, the CPLP Cultural Television Channel will be able to start exporting a great quantity of services and videos to Africa, Europe and South America. (See financial projections)

The economic feasibility of the project will occur with the commercialisation of several products, like:

1.   The sale and production of cultural programmes to various institutions like:
-       BBC
-       TV Culture Brazil
-       RTP
-       Discovery
Etc.

All over the world commercial networks for television cultural programmes have been created. Several television channels in different countries have regularly acquired programmes by producers spread out all over the world. Not only, several international organisations, like the World Bank and the United Nations, including FAO and Unesco, among many others, need programmes for public education, like programmes oriented to alert and to educate people concerning endemic and epidemic diseases. AIDS, malaria, tuberculosis are a very few examples.

These world institutions have made a great effort to develop television programmes with humanitarian objectives. But, in Africa – undoubtedly the continent most needed of such programmes – there is no television station, in present times, with capability to make face to such a need.

Therefore, the television teams that had been responsible for such programmes are, practically in its totality, placed in countries of the called First World, strange to the local population's concrete reality.

Programmes focusing new agricultural techniques or even oriented to agricultural, commercial and industrial education are essential elements of such a repressed demand.

The price for each institutional campaign, with an average of 10 films produced per campaign, is of about USD.15.000$ and the capacity of the CPLP Cultural Channel will be as follows:

First Group* - from 2 to 4 campaigns per month.
Second Group – from 1 campaign in two months to 1 campaign per month

This represents:
First Group   a potential annual income of about US$.540.000
Second Group a potential annual income of about US$.90.000

*First Group: Angola, Cape Verde, and Mozambique. Second Group: Guinea-Bissau, Sao Tome e Principe, East Timor (Lorosae). Reference Group: Brazil, Portugal.

Obs. For the case of Guinea-Bissau and East Timor, it is expected a decrease of such incomes for US$.216.000 and US$.36.000.

2. The sale and production of no cultural programmes to other television channels, like:
-       series (novels)
-       talk-shows Etc.

Many countries of the region need to produce television programmes, but do not have capability to do it. Thus, they seem themselves obliged to search expensive productions in Europe. With only one programme sold-three daily hours-the channels of the First Group* would receive in incomes the equivalent of about two times of the whole investment. The Second Group countries will not have conditions at the beginning to product cultural programmes to other television channels.

*First Group: Angola, Cape Verde, and Mozambique. Second Grope: Guinea-Bissau, Sao Tome e Principe, East Timor (Lorosae). Reference Group: Brazil, Portugal.

The price for this type of television programme is:

Auditory programs   US$.12.500 per programme
Talk-shows   US$.4.000 per programme
Interviews   US$.4.000 per programme

The capacity of production, of these programmes by the countries of the First Group will be:

In the FIRST PHASE:
Talk-shows   2 per month
Interviews   4 per month
In the FINAL PHASE:
Auditory programmes   6 per month
Talk-shows   20 per month
Interviews   20 per month

Thus, the potential income will be:

FIRST PHASE
Talk-shows   US$.96.000 per year
Interviews   US$.192.000 per year

FINAL PHASE
Auditory programmes   US$.900.000 per year
Talk-shows   US$.960.000 per year
Interviews   US$.960.000 per year

Obs. All values must be decreased for Guinea-Bissau and East Timor in the first years, as it is referred in the detailed feasibility studies.

The price of the novels is much higher. For each complete novel, with about 120 chapters, the price is about US$.2.500.000 and the capacity of the First Group of the CPLP Cultural Channel (in the final phase) will be of one novel per year.

3. There is the possibility of commercial use of the transmission time beyond the six hours reserved to culture and education.

The conventional day in television is of 18 hours, of which only 6 hours would be no commercial. We would have, therefore, 12 hours of no cultural transmission, which should be freely commercialised.
Even the period of six hours- divided into two sections of three hours, the first one dedicated to culture and second section of three hours to education-has a great potential for sponsoring.

Each commercial hour of transmission can include up to 12 minutes of advertising. The price for each 30 seconds of advertising is:

Noble time   from 6PM to 11PM   a b o u t US$.400 each 30"
Normal time   rest   about US$.200 each 30"

Concentrating the commercial programming in the noble time. With 4 hours of transmission in this period, we would have:

Educational television: from 7AM to 9AM
From 3PM to 4PM
Cultural television: from 4PM to 7PM
Commercial television: from 9AM to 3PM
From 7PM to 1AM

Thus, the commercial period would comprehend 6 hours in normal time, 4 hours in noble time and 2 hours in normal time again.

Begin 12 minutes per hour for advertising, the commercial period could have:

Noble time   48 minutes   24   films   US$.9.600
per day
Normal time   96 minutes   48   films   US$.9.600
per day
Total of the potential income........... US$.19.200
per day

                or US$.576.000 per month
                or US$.6.912.000 per year

It is believed that in the first phase of the project (after one year), the CPLP Cultural Television Channels of the First Group will be able to start with an income from commercial advertisement of about US$.1.000.000 per year-value which is predicted to increase in the follow months.

4. The rent of the studios to television teams of other countries can be another source of incomes. Many producers who cover events in Africa need to move many times to their countries of origin during the video works, because there is not technical support in Africa. The same phenomenon happens with the cinematography and the journalism productions. It is not difficult to imagine, for example, the serious problems journalism teams have had, for example, with essential components like batteries, lighting, electronic components etc., which only can be easily find in Europe.

The period of renting of a television studio is of 12 hours. Each period has price of about US.4.000. After the first year, the studio should be rented for 12 hours per each period of 3 days. So, the rent per month, in this period (First Phase), will be able to generate incomes of about US$.120.000 per month in the First Group.

5. Support services to other television networks in all areas, including novels, mini series, docdramas series etc.

Many countries, principally in Africa, do not have technical conditions to develop this kind of programmes, but they have a strong internal repressed demand in this sense.

Each team, abroad, has a price of US$.1.000 per day

(two people), after the costs of dislocation, hotel and meals. In the first months (first phase) it will be organised one team for works in other countries. In the final phase it is predicted to have up to 5 teams with such function. The capacity of each team is of about 20 days per month. Therefore, the support to other television networks-when took in its full potentiality-will be able to represent up to 20 days per month in the first phase. This represents a potential income of about US$.240.000 per year, and 100 days per month of works in the final phase, signifying about US$.1.200.000 per year of incomes, always referring to the First Group countries.

6. Colloquies, seminars, videoconferences and meetings of different natures.

The building of the CPLP Cultural Television Network will have a medium size auditorium, (First Group), with all conditions to receive seminars, colloquies and meetings of the most diverse nature. Such seminars, videoconferences etc. are important not only for the increase of the incomes, but also to attract specialists of the most different areas, being an important element for the development of all region as well as for the diffusion of the local, regional and continental cultural values.

The meetings, seminars and colloquies attract, in average, about 300 people per event. The price-excepting meals, hotels and transportation-per each participant is of about US$.50 per day. It will be a capacity for up to four events of this type per month, what could represent an income of about US$.60.000 per month or US$.720.000 per year (final phase-after 5 to 7 years).

The project should also turn possible.
-        classes of professional formation in diverse disciplines
-        support for a multimedia high technology centre (First Group)
-        support for a multimedia high cultural centre (Final Group)

GUINEA BISSAU TELEVISION NETWORK

PROFITABILITY & CASH FLOW

CHAPTER - 8

PROFITABILITY AND CASH FLOW

8.1   Estimation of cost of production and working results for the first five years of operation.

The estimated cost of production of working results

for the first five years of operation are given in the chapter "Financial Projection".

8.2   Cash flow statement for the company as a whole, for five operating years of the project based on the estimates of working results.

A detailed cash flow statement for the company as a whole for five operating years is given in the chapter of "Financial Projection".

8.3   Projected balance sheet for five operating years for the company as a whole.

The balance sheet for five operation years for the company as a whole is given in the chapter of "Financial Projection".

GUINEA BISSAU TELEVISION NETWORK

ASSUMPTIONS

CHAPTER - 9

ASSUMPTIONS

1.   The Guinea Bissau Television Network shall generate the income from the following sources.

a)   Sale of cultural programmes.
b)   Sale of commercial programmes.
c)   Sale of advertisement time during transmission of TV programmes

d)   Hiring of studios.
e)   Supply of technical services
f)   Hiring of conference hall

The detailed assumptions for each of the above mentioned activities are as follows:

a)   Sale of cultural programs:

I.   It is assumed that four cultural programs shall be produced per month in the first year, which can be sold to other countries. This figure will increase to 5 programs per month in the second year, six programs per month in the third year and so on.

II.   It is assumed that the selling price shall be US$.60 per program.

b)   Sale of commercial programs:

I.   The commercial programs consist of talk shows, interviews and auditory programs.

II.   It is assumed that two talk shows shall be produced per month in the first year, three talk shows shall be produced in the third year, four talk shows per month shall be produced in the third year and

so on.

III.   It is assumed that the selling price of one talk show program shall

be US$.1600.

IV.   It is assumed that four interviews shall be produced per month in the first year, five interviews shall be produced in the third year, six interviews per month shall be produced in the third year and so on.

V.   It is assumed that the selling price of one interviews program shall be US$.1600.

VI.   It is assumed that no auditory programs shall be produced in the first and the second year. Only when the people have two years of experience then in the third year one auditory program shall be produced per month. In the fourth year two programs shall be produced per month and so on.

VII.   It is assumed that the selling price of one auditory program shall be UD$.5.000.

c) Sale of advertisement time during the transmission of T.V. programs:

I.   There shall be a total transmission of 8 hours per day in the first year. It will go on increasing by two hours in the second and subsequent years.

II.   Out of which advertisement shall be available for programs with transmission period of four hours in the first year. The programs in

which advertisements could be available shall increase by two hours per year in the second and subsequent years.

III.   In one hour T.V. transmission, 12 minutes of advertisement shall be allowed.

IV.   Out of total TV time 33% time shall be considered as prime time and remaining 67% shall be considered as non prime time, in the first, second and third years. From the fourth year onwards the prime-time advertisements shall remain constant.

V.   The advertisement rates shall be US$ 320 for one minute of prime time advertisement and US$ 140 for one minute of non prime time advertisement.

d)   Hiring of studios:

I.   It is assumed that the studios shall be taken on hire for a shift of 12 hours, two times in a month in the first year. It will increase to

three times in a month in the second year.
II.   The rent per day shall be US$.1.600.

e)   Supply of technical services:-

I.   It is assumed that one team, consisting of two technically qualified people shall be available in the first year. In the second year two

such teams shall be available. In the third year three such teams shall be available and so on.
II.   It is assumed that in the first year the team shall be hired for 10 days in a month. In the second year the teams shall be hired for 12

days in a month. In the third year the teams shall be hired for 14 days in a month and so on.

III.   The rate of hire shall be US$.400 per team per day.

f)   Hiring of conference hall :-

I.   It is assumed that the conference hall shall be taken on hire for two times in a month in the first year, three times in a month in the

second year and four times in a month in the third year and so on.

II.   The hire charges shall be US$.200 per conference.

2.   The unit is planning transmission of three hours of educational programs, three hours of cultural programs and two hours of commercial programs per day in the first year. Thus, there shall be a total transmission of eight hours of programs per day in the first year. The total transmission hours shall increase by two hours per day in the second and subsequent years. Thus, in the second year there shall be total transmission of 10 hours per day and in the third year there shall be a total transmission of 12 hours per day. The cost of software production is assumed at US$.1.000 per hour.

3.   The cost of miscellaneous consumable items is assumed at US$.14.400 in the first year. This will increase by 10% in the second and subse-

quent years.

4.   The cost of stores and spares consumed is assumed at US$.16.800 in the first year. It will increase by 10% in the second and subsequent years.

5.   The cost of repairs and maintenance is assumed at 0.1% of the total cost of plant and machinery in the first year. This cost will increase up to 0,3% in the fifth year.

6.   It is assumed that the unit shall use 1.000 units of power every day or 365.000 units of power in the first year. The consumption of power shall increase by 10% in the second and subsequent years. The cost of power is assumed at 2 cents for one unit.

7.   It is assumed that the unit shall consume 1.000 litters of diesel per month. The cost is assumed at 13 cents for one litter. It will increase by 10% in second and subsequent years.

8.   The unit shall have a technical staff of 35 people and administrative staff of 15 people.

9.   Depreciation is calculated on reducing balance method. The rate of depreciation is taken at 10% for building, 25% for plant and machinery, 10% for furniture and fixtures and 20% on vehicles.

10. Advertisement expenditure is assumed at 0.2% of the sales.

11. It is assumed that the stock of raw material and stores shall be 30 days of consumption, receivables shall be 45 days sales and creditors shall be 15 days purchases.

12. The term loan shall be for a total period of 25 years. It shall have a moratorium of five years. It shall be repayable in 20 equal yearly instalments.

PROJECT REPORT

OF

MOZAMBIQUE CULTURAL TELEVISION NETWORK

AT

MOZAMBIQUE

AFRICA

PROMOTERS

REPUBLIC OF MOZAMBIQUE

AFRICA

TELEVISÃO SONHO

Rede Planetária de Televisão Cultural para a Paz Mundial

MOZAMBIQUE TELEVISION NETWORK

INTRODUCTION

CHAPTER - 1

INTRODUCTION

1.1   Description of the organisation through which this project is being taken up.

The CPLP is an international organisation, created in 17 of July of 1996, with Headquarters in Lisbon and consisting of the following State Members: Angola, Brazil, Cape-Verde, Guinea-Bissau, Mozambique, Portugal, Sao Tome e Principe and East Timor (Timor Lorosae).

The CPLP has as its objective the politic-diplomatic relationships between its Members, mainly referring to the United Nations and the World Bank, the relationship EU/MERCOSUL and the accomplishment of the Europe-Africa Summit.

The CPLP is also dedicated to the co-operation, particularly in the economic, social, cultural, legal and technical-scientific matters; as well as to the projects of promotion and broadcasting of the Por-

tuguese Language, nominated to the improvement of the International Institute of the Portuguese Language and the creation of a Bibliographical Fund.

All its decisions are taken by consensus.

2.2   Description of the main reasons for setting up an African Cultural Television Network.

Africa is a continent submerged into dramatic problems. All these humanitarian problems have in the education and culture their most direct and objective root.

The African countries speaking Portuguese are among the African countries with the lowest level of development.

The whole situation is so dramatic that the implementation of the CPLP Cultural Television Network is more than urgent.

It is hoped that with this Television Network the people shall have better understanding of various diseases and other problems of life. e.g. AIDS is prevalent disease in Africa. But as the people are illiterate the information about the disease cannot be spread by any other method. In such cases the Television Network becomes a very effective audio-visual means of communication.

The project also intends to be a base of a massive production of health campaigns, specially to reinforce the combat against the AIDS, the infect-contagious diseases- endemic or epidemic-through a process of global education of the local and regional populations.

The CPLP Cultural Television Network is a project for peace and its objective is not to compete with local television channels already installed, but to interchange with them, promoting a future global web dedicated to knowledge and development.

1.03 Description of the main objects of the project.

The major objective of the project is to assist the CPLP countries to develop the first African television network channel oriented to cultural and educational themes with the follow orientation:

- diffusion of the Portuguese language
- education in all areas
- reinforcement of the local and regional cultures
- formation of technical personal on communication and tele-education
- formation and information for a better participation of the CPLP nations in the process of global development
- to support the implantation of the Institute

of Portuguese Language

The development of such a project will create an irradiation pole from the countries with the project implementation to the whole African continent, constituting an African network of research and culture, discovering and valorising different cultural aspects and, consequently, reinforcing identity factors.

With the stronger cultural identity, it is expected to have lower levels of violence in social behaviour as well as a relevant increase of the educational performance in all its sectors.

Systematic educational objectives are complementary to the general orientation of this project, which is cultural in its fundament.
A very large positive impact is also to be expected with respect to several sectors of the economy, greatly improving the overall quality of services.

1.4   Description of the mission status of the project.

The mission of the project is:

-        to rescue lost values and cultural information
-        to reinforce historical elements

- to rescue old artworks
- to give a new value to new artworks
- to communicate practical and objective information on health
- to communicate practical and objective information on agriculture
- to communicate practical and objective information on economy

1.05 Description of the philosophy of the project.

The CPLP Cultural Television Network will have in its philosophy, as a support to its mission, the follow fundaments:

- to work with local people
- be focused on quality
- intercommunication

Three words can define the CPLP Cultural Television Network philosophy:
People, quality and intercommunication

Following these three key words of its philosophy, a single sentence also defines its mission: "A Continent of Culture".

Two main sectors characterise its global structure:
1. Cultural Sector
1.1.   To develop programmes on:
-       Music

- Architecture (also vernacular)
- Archaeology
- Literature and poetry
- History
- Tourism
- Dance
- Plastic arts
- Astronomy
- Science in general
- gastronomy
- health
- anthropology
- local cultures
- others

1.2   The programmes can be of all types, the most important is to be attentive to communication. All programmes must reach the highest possible number of spectators.

2. Educational sector.
2.1   Systematic education in short practical courses.
2.2   Educational series with practical and objective information on:
- health care
- family
- agriculture
- managing
- services
- civil construction

As the CPLP Cultural Television Network must not be supported by the respective governments in the future, it will have a daily time of free commercial programmes up to 12 hours. This is a good solution to pay its costs and support its development.

1.06 Description the magnitude of the project.

The project proposes a Television transmission time of eight hours per day in the first year. It will increase by two hours per day in the second and subsequent years.

The project also envisages training of local people in all the related areas of Television. It is proposed to develop a local pool of talent which can be utilised not only in this particular country but also in other countries of Africa.

The total cost of the project of the Mozambique Cultural Television Network is US$.18.400.000. Out of this it is proposed to obtain a soft loan of US$.15.400.000. The remaining amount of US$.3.000.000 shall be contributed by the Government of Mozambique in terrain for the installation of the respective building, infrastructures and logistic support.

For the purpose of the Mozambique Cultural Television Network a company shall be formed with the participation of the Government. The Contribution of the Government in the form of share capital shall be US$.3.000.000.

The term loan shall be on soft terms. The total period of the loan shall be 23 years. Out of which five years shall be a period of moratorium. The entire loan shall be repaid in the remaining 17 years, in 17 equal yearly instalments.

The Government of Mozambique will give the sovereign guarantee for the referred loan.

MOZAMBIQUE TELEVISION NETWORK

THE COUNTRY

CHAPTER - 2

THE COUNTRY

2.1 - Description of the level of development of Mozambique.

Before the peace accord of October 1992, Mozambique had been devastated by civil war and was one of the poorest countries on the globe.

Prospects subsequently improved, and with its solid economic performance in 1996-97, Mozambique has begun to exploit its agricultural, hydropower, and transportation resources. Foreign assistance programs help supply the foreign exchange required to support the budget and pay for imports of goods and services.

The restoration of electrical transmission lines to South Africa and the completion of a new transmission line to Zimbabwe (permitting the giant Cahora Bassa hydropower plant to export large amounts of electricity), proposed construction of a natural gas pipeline to South Africa, and reform of transportation services will greatly improve foreign exchange receipts. The Mozambique and South African Governments are developing the Maputo corridor, linking the port of Maputo with Witbank, South Africa. In the past few years, more than 700 state enterprises have been privatized, including the country's largest commercial bank and a number of sizeable manufacturing firms. Other pending reform measures are the reform of tax collection and the facilitation of private enterprise in the transportation, energy, and telecommunications sectors.

OFFICIAL NAME: Republic of Mozambique
CAPITAL: Maputo
SYSTEM OF GOVERNMENT: Unitary Multiparty Republic
AREA: 798,800 Sq Km (308,418 Sq Mi)
ESTIMATED 2000 POPULATION: 17,857,100

LOCATION & GEOGRAPHY: Mozambique is located on the south-east coast of Africa. It is bound by Swaziland to the south, South Africa to the southwest, Zimbabwe to the west, Zambia and Malawi to the north-west, Tanzania to the north and the Indian Ocean to the east. The country is divided into two topographical regions by the Zambezi River. (1.) North of the Zambezi river, the narrow coastline moves inland to hills and low plateaux, and further west to rugged highlands, which include the Livingstone-Nyasa Highlands, Namuli or Shire

Highlands, Angonia Highlands, Tete Highlands and the Maconde Plateau. (2.) South of the Zambezi River, the lowlands are broader with the Mashona-land Plateau and Lebomo Mountains located in the deep south. The country is drained by five principal rivers and several smaller ones with the largest and most important the Zambezi. The country has three lakes, Lake Nyasa or Malawi, Lake Chiuta and Lake Shirwa, all in the north. Major Cities (pop. est.); Maputo 931,600, Beira 298,800, Nampula 250,500 (1991). Land Use; forested 18%, pastures 56%, agri-cultural-cultivated 4%, other 22% (1993).

CLIMATE: Mozambique has a tropical climate with two seasons. A wet season from October to March and a dry season from April to September. Climatic conditions, however, vary depending on altitude. Rainfall is heavy along the coast and decreases in the north and south. Annual precipitation varies from 500 to 900 mm (20 to 35 inches) depending on the region with an average of 590 mm (23 inches). Cyclones are also common during the wet season. Average temperature ranges in Maputo are from 13 to 24 degrees Celsius (55 to 75 degrees Fahrenheit) in July to 22 to 31 degrees Celsius (72 to 88 degrees Fahrenheit) in February.

PEOPLE: The majority of the population belong to local tribal groups which include the Makua-Lomwe who account for 37% of the population while the Shona account for 10% and the Tsonga for 23%.

Other ethnic minorities include Europeans, mainly Portuguese, Euro-Asians and Indians.

DEMOGRAPHIC/VITAL STATISTICS: Density; 18 persons per sq km (47 persons per sq mi) (1991). Urban-Rural; 13.2% urban, 86.8% rural (1980). Sex Distribution; 49.3% male, 50.7% female (1990). Life Expectancy at Birth; 44.9 years male, 48.1 years female (1990). Age Breakdown; 44% under 15, 26% 15 to 29, 16% 30 to 44, 9% 45 to 59, 4% 60 to 74, 1% 75 and over (1990). Birth Rate; 45.0 per 1,000 (1990). Death Rate; 18.5 per 1,000 (1990) . Increase Rate; 26.5 per 1,000 (1990). Infant Mortality Rate; 141.0 per 1,000 live births (1990).

RELIGIONS: Around 48% of the population follow local native tribal beliefs while 39% are Christians and 13% are Muslims.

LANGUAGES: The official language is Portuguese which is used for government, education and commerce purposes. A variety of local tribal languages are also widely spoken.

EDUCATION: Aged 25 or over and having attained: no formal schooling 80.7%, primary 18.2%, secondary 0.9%, higher 0.2% (1980). Literacy; literate population aged 15 or over 32.9% (1990).

CURRENCY: The official currency is the Metical (MT) (plural; Meticais) divided into 100 Centavos.

ECONOMY: Gross National Product; USD $1,375,000,000 (1993). Public Debt; USD $4,650,000,000 (1993). Imports; USD $1,018,000,000 (1994). Exports; USD $149,500,000 (1994). Tourism Receipts; N/A. Balance of Trade; USD - $868,000,000 (1994). Economically Active Population; 5,671,290 or 48.6% of total population (1980). Unemployed; 1.7% (1980).

MAIN TRADING PARTNERS: Its main trading partners are the USA, Portugal, France, Iraq, Japan and Singapore.

MAIN PRIMARY PRODUCTS : Bananas, Bauxite, Cashew Nuts, Cassava, Cereals, Coal, Coconuts, Copper.

MAJOR INDUSTRIES: Agriculture, Chemicals, Cement, Food Processing,

Mining, Petroleum Products, Textiles.

MAIN EXPORTS: Cashew Nuts, Cotton, Shrimp, Tea, Textiles.

TRANSPORT: Railroads; route length 3,271 km (2,033 mi) (1988), passenger-km 75,300,000 (46,789,000 passenger- mi) (1988), cargo ton- km 231,800,000 (158,760,000 short ton-mi) (1988). Roads; length 26,095 km (16,215 mi) (1989). Vehicles; cars 84,000 (1989), trucks and buses 24,000 (1989). Merchant Marine; vessels 114 (1990), dead weight tonnage 29,153 (1990). Air Transport; passenger-km 553,829,000 (344,133,000 passenger-mi) (1990), cargo ton-km 60,636,000 (41,530,000 short ton-mi) (1990).

COMMUNICATIONS: Daily Newspapers; total of 2 with a total circulation of 81,000 (1994). Radio; receivers 620,000 (1994). Television; receivers 35,000 (1994). Telephones; units 62,100 (1993).

MILITARY: 50,000 (1993) total active duty personnel with 90.0% army, 2.2% navy and 8.0% air force while military expenditure accounts for 7.6% (1993) of the Gross National Product (GNP).

MOZAMBIQUE TELEVISION NETWORK

THE PROJECT

CHAPTER - 3

THE PROJECT

3.1 Description of the initial steps to be taken for the implementation of the project.

The first objective of the CPLP Cultural Television Network is to install its structures in the following countries:

- Sao Tome e Principe
- Angola
- Mozambique
- Guinea-Bissau
- Mozambique
- Timor

Thus, these six countries will structure the basic network of the whole project.

Mozambique, Mozambique and Angola will form a FIRST GROUP; Sao Tome e Principe, Guinea-Bissau and East Timor will form a SECOND GROUP; and, finally, Brazil and Portugal will form a called REFERENCE GROUP.

However, the project will start at the same time in all countries. A more precise description of the meaning of each group is made below.

The basic physical components of the project are:

1.   The creation of companies, in each of the States of the CPLP, oriented to the objectives above referred.
2.   The State of each country – who will be the responsible and guarantee of the payment of the financing – will also have a participation in the companies.
3.   All strategies will be oriented to lowest costs in long term.
4.   The whole project will be developed by phases.
5.   The timetable of transmission will be defined case by case.
6.   The project will include:
6.1.   a complete plan for an equipment network.
6.2.   a complete plan for a software network.
6.3.   all procurements
6.4.   equipment transportation and installation
6.5.   all configurations
6.6.   creation of the infographic system

6.7.   technical formation of the personal

6.8.   architecture and engineer projects for the building

6.9.   construction of the building organised in phases

1.2   Description of the procedure by which the necessary programmes will be produced for transmission on the channel.

The television programs are known as software in this field. The unit is planning transmission of three hours of educational programs, three hours of cultural programs and two hours of commercial programs per day in the first year. Thus, there shall be a total transmission of eight hours of programs per day in the first year. The total transmission hours shall increase by two hours per day in the second and subsequent years. Thus, in the second year there shall be total transmission of 10 hours per day and in the third year there shall be a total transmission of 12 hours per day. The cost of software production is assumed at US$.2.500 per hour. The amount of US$.2.500 per hour is taken based on cost of production of such programmes in Portugal.

To contract a similar system from other television channel, only for one country, for a transmission period of 6 hours daily, the cost in, market (eg. Portugal*) is of about US$.500.000 per month. This cost represents a total of about US$.6.000.000 per year. In 16 years, excluding the first year of implementation, the costs for contract the six daily hours of transmission from other television channel, in present market price, is of about US$.100.000.000. The six countries would have an estimated cost of about US$.600.000.000 in a 16 years period of time.

These costs are based on the prices presently used in Portugal, as for example the programme Hora Viva – Seguranca em Directo, transmitted every day, excluding weekends, from 7 to 10 AM.

Another example is the channel Canal Noticias Lisboa CNL. Only in its first year of installation, it was expended about US$.10.000.000, and the equipment as well as the building were not property of the channel, but rented.

In fact, however the project will be implemented by phases, after 5 to 10 years of regular work, it is predicted to have reached a similar value in incomes as the examples above.

*value took in the Portuguese market, below the average of the European prices in about 30%.

The first ten months of works will be oriented to:
-	design of the equipment and software structures

- start the first productions
- elaboration of the projects of architecture and engineering
- procurement and acquisition of all equipment
- construction and supervision of the works
- installation and configuration of the equipment

So, the first year must be used to the implantation of the project. However, as to accelerate the chronogram, some operations should be started before this timing, temporarily installed in a different building.

The first phase will be oriented to:
- video productions (external and internal)
- edition
- elaboration of cost programmes

3.3   Description of the requirements of Power, Fuel and Water. POWER
It is assumed that the unit shall use one thousand units of power of everyday. Thus, the unit shall use 365.000 units of power in a year. The costs of power is assumed at 3.5 cents for one unit. The consumption of power shall increase by 10% every year.

FUEL

The unit shall use D.G. set for generation of electricity, when there is an interruption in the supply. The hours of disruption in the normal supply can not be predicted accurately. Hence, the hours for which the D.G. sets shall be used also can not be predicted accurately. However, it is assumed that the unit shall utilise one thousand litters of diesel per month. The cost is assumed that 23 cents for one litter. Hence, the total cost for one thousand litters shall be US$.230. The consumption of diesel shall increase by 10% every year.

WATER

The unit does not need water for fixed commercial operation. However, it shall have a strength of 50 people as employees. Water shall be required for drinking and sanitation purposes. It is assumed that 3000 litters of water shall be required everyday. The cost of obtaining this water is assumed at US$.7 everyday. The consumption of water shall increase by 10% every year.

3.4   Description of the manpower requirement of the project and description of the steps taken by the company to train the manpower.

The total staff strength shall be 17 people in the first year.

The system of modern TV Station is designed to achieve sustained operating efficiency and transmission. This, of course, entails a certain degree of sophistication in the production and transmission. However, considering the socio-economic situation in Mozambique, a reasonable balance has to be struck to obtain optimum performance and at the same time create gainful employment. While working out the manpower requirement for this project to be kept on direct rolls of the company totalling 50 staff and operators, the above consideration have been kept in mind.

Manpower requirement

The direct manpower required for the proposed Unit is about US$.1.260.000 per year. The manpower requirement as indicated in this chapter has been planned keeping in view the following guidelines:

Effective co-ordination among the various departments. Judicious distribution of responsibilities. Capacity utilisation of the TV Station with optimum manpower.

Details of manpower requirement for the TV Station is given in the financial section.

In line with the prevailing practice, the Security guards, office peons and unskilled labour etc., are normally employed on contractual/casual basis, and their cost has to be included in the Factory overheads. However in this report provision has been made to employ the above staff and their salaried have been included.

The various departments proposed shall be under the direct responsibility of the Station Director. He shall be assisted by Deputy Directors, who will look after the complete TV Station and its various day to day activities and Marketing. They shall be responsible for achieving the envisaged targets and sales forecasts. They shall be assisted by a team of Managers from production, marketing and accounts.

To run this project a labour force which shall usually be composed of unskilled and skilled workers shall be employed. The first are those who do not undergo any king of specific training or education, while the latter have to do so in order to master their jobs.

When evaluating an investment project from an employment point of view, its impact on both unskilled and skilled labour has been taken into account. Not only direct employment, but also indirect employ-

ment has been considered. Direct employment refers to the new employment opportunities created
within the project; indirect employment concerns
job opportunities created in other projects linked
with the project which is being formulated.

The implementation of large and sophisticated
projects generally contributes to the development
of local skills and capabilities in a country. Furthermore, they help to change traditional values, attitudes and the behaviour of the society, to build
up an enterprising spirit among the people, to develop a desire for changing and improving the existing conditions of life, to introduce better work
discipline and thus to change the very pattern and
basis of economic development. The TV industry
is already well established in Africa. Location of TV
industry activities in Africa, has certain favourable
factors and advantages, in setting up this type of industry, as availability of acceptable level of education among the supervisory staff adequate technical and managerial skills developed over a long period, and availability of cheap labour are assured.

The organisation structure will vary from TV Station to TV Station in the industry and as such the
pattern proposed herein can be considered only as
suggestive and provisional.

A well-knit organisation structure headed by a Station Director and Manager, with the supporting

staff will be developed progressively during project implementation. Soon after the plant becomes operative, a good number of project staff will be absorbed in the organisation. Certain additional staff also get added to ensure smooth and efficient management of the operating unit.

The Manager Production will have a degree in communication and will be accountable and answerable to the Station Director for all TV operations including planning, production, material management, TV utilities, quality control, production cost and budgetary control, TV safety, discipline and layout relations. His main function should be to ensure achievement for quality and quantity targets of production at reasonable cost and should constantly strive to improve TV performance. In the discharge of his multifarious duties and responsibilities he will be assisted by supervisors and adequate staff for day to day activities.

His duties are to ensure that targets are maintained through effective utilisation of 3 M's viz: Machinery, Men and Materials. These targets should be translated in terms of targets for individuals under him such as supervisors, skilled workers etc. He has to ensure that the equipment at his disposal gets prompt attention on breakdown is adhered to strictly. He has to further ensure that there is a strict quality control exercised over raw materials also. He will be assisted by Deputy Manager etc. who and will report to the General Manager.

GENERAL MANAGER (Comm.) He will report to the Station Director. His main duties will be prepare sales forecasts and budgets, study and monitor the export market for the company's products and advice on ways and means of increasing sales. He will also ensure that consumer complains are solved in an appropriate manner. He will also ensure that the right materials are available for production at the right time.

He will be responsible for all aspects connected with the export procedures and keep the management updated on all matters relating to the Govt. policies on polymers and Exports and above all the world market.

COMPANY SECRETARY & FINANCIAL CONTROLLER. He will report to the Station Director and will be responsible for the departments of administration, financial planning, budgetary control, cost accounting, tax management payroll accounting, etc. and all personnel matters. He will be assisted in their duties by their respective assistants to assist in day to day activities.

Manpower planning and production

The central issue here may be one of scale. Production is staffed bye personnel, but the process may be labour-intensive or capital-intensive. Either way, planning will include:

1.   Analysis of labour supply and demand factors in relation to skills and training needs.
2.   Procedure for manpower recruitment, together with selection processes.
3.   The formation of industrial relations policies necessary for effective work place bargaining, disciplinary measures and dismissal procedures.
4.   Analysis of the effective use of human resources.
5.   Conditions necessary to maintain adequate levels of motivation. The scale of the problem is likely to be directly proportional to the method of production.

The availability of main persons is not going to be easy since TV industry is currently in its infancy in Africa.

Training needs

The selection and training of the required manpower for the proposed project has to be planned in advance.

The key personnel should be selected and trained suitably. The training would be carried out in the following manner:

Basic training on the concept of TV industry before construction begins with visits to similar TV Stations in other countries. On site training during the construction phase of the project. On job training during the commissioning phase of the project. On job training during operation of the TV immediately after commissioning. The training of the key personnel such as Station Director should be carried out in all the phases. The training of other operating personnel should be suitably carried out during the construction, erection and operation phases in addition to training them by visits to similar plants operating in other countries.

Besides training the key operating staff described above, in TV training should also be given to other employees at skilled operating level to enable them to understand the process equipment in the project and prepare them to operate a maintain their respective sections safely, efficiently and skilfully. The above training should be carried out during construction, commissioning and operating phases of the project.

Training is necessary in order to enable personnel to acquire the skills and knowledge necessary to perform a task to an acceptable standard. The length of the training period and training methods will, of course, vary from job to job. Training is essentially a learning process, and in order that progress can be successfully monitored certain conditions are necessary.

1.   The training needs of both the individual and the organisation shall be identified and analysed.

2.   Targets and standards shall be set for the trainee, which are within his capabilities.

3.   The pace of the training programme should reflect the trainee's ability to maintain progress in properly absorbing the same.

4.   The trainee shall receive regular feedback of results. Any problem areas shall be highlighted, discussed and resolved.

5.   As the trainee progresses the amount of information provided shall be gradually reduced, thus inducing a feeling of independence and competence.

It is common place to find a wide variety of tasks in an organisation and each will require varying degrees of skill, effort and responsibility. This being so, it is inevitable that rates of pay will also vary and the differentials between the jobs will reflect their relative values. However, other factors such as local market conditions, bargaining strengths and traditions also influence a company's payment structure and a great deal of planning is required if rationalisation is to be achieved. One technique which has been successfully adopted by many companies to establish an equitable wage structure is job evaluation. In a job evaluation exercise a comparison is made of common criteria over a range of jobs, and the resulting analysis may be linked to a points allocation or job ranking system, and hence to a wage scale.

In conducting a job evaluation exercise it is important to cover a reasonable variety of tasks within the whole spectrum. For each, a job description is prepared setting out details of the duties and responsibilities undertaken by the employee together with a statement about his working conditions. Very often this task is undertaken by work study personnel since, they are responsible for determining methods of operation and evaluating the work content of the job. Each job is assessed factor by factor, resulting in a comprehensive comparative analysis.

The individual is the most important resources of any company and only people who are well trained, well motivated and adequately rewarded will provide a positive and synergistic contribution towards the company's objective and its organisational health.

In most cases, the factors, which may be weighted according to relative value, are as follows:

1.   Skill-education, experience and training.

2.   Effort-both physical and mental.
3.   Responsibility-for equipment, materials, initiative etc.
4.   Working conditions-general conditions, risk of accident and injury.

Pay policies affect not only individual employees but the whole organisation, and the rewards and objectives vary at different levels within the enterprise.

Industrial relations

An effective industrial relations policy is important, since is the system through which employees take part in decision-making and in many instances it affects the while atmosphere of employer/employee relationships. An industrial relations policy is essentially a set of rules whose determine procedures for negotiation on such matters as:

1.   Wage and salary scales.
2.   Terms and conditions of work.
3.   Disputes and grievances.
4.   Recruitment and dismissal.
5.   Other issues of mutual interest, e.g. closed shop, redundancies and joint consultation.

In order to promote an atmosphere of co-operation, and to minimise conflict, the needs of management and work people must be recognised by both sides. Trade unions exit to protect the interests of their members and improve their working conditions. Management, while aware of the pressures and constraints imposed by the trade unions, have a duty to maximise the use of resources at their disposal, which may be expressed in relation to profitability, return on investment, level of service, sales volume, market share and cost-effectiveness. The strategies adopted in attempting to solve industrial relations problems will vary from company to company, and indeed from union to union, but there is no doubt that they will be influenced by both internal and external factors.

Internal factors

1.   The attitudes of employees to management, and management to employees.
2.   The leadership style of management.
3.   The bargaining strength of both parties.
4.   The number of negotiating bodies.
5.   The prosperity of the company.

External factors

1.   The extent to which parent boards influence company management, and district officials influence or control local shop stewards.
2.   Whether or not bargaining is conducted at plant, local or national level.

3.   Government policy towards industrial relations.

4.   The economic situations nationally, locally or within the company itself.

3.5   Describe the project implementation schedule.

Implementation of this project is a challenging task and calls for meticulous planning, scheduling and monitoring to realise the project goals within the budgeted cost and time frame. The goal can be achieved by adopting modern project management techniques.

To implement this project adequately, a team of engineers and project personnel having requisite education and experience are being appointed, to whom a detailed Work Breakdown Structure (WBS) in a logical order of activities, shall be supplied shortly, keeping in view cost estimation, scheduling, and to help monitor and control of the project. It is proposed to be formulated in conjunction with the objectives of each activity and goal settings. Project shall be programmed and controlled by network analysis techniques. Before the application of the network analysis techniques, the project personnel shall be acquainted with their capabilities in saving time, resources and costs. The training of project personnel and engineers at levels shall be provided for proper control of project progress and taking of timely corrective actions to re-align these efforts to meet ore stated objectives. It is proposed to gear programming and control system, i.e. project implementation system, which will ensure an integrated approach to project implementation. Project management activities shall be determined in advance and all activities carried out be project personnel as well as those to be contracted shall be identified.

Responsibility for project implementation shall be clearly defined. The forms of project organisation range from project oriented to functional organisation, while most of the cases are combinations of the two, with certain adaptations to prevailing conditions. It is impossible to over emphasise the importance of establishing a team of a task force for implementing the project with a designated leader to co-ordinate and guide its functions.

Project manager: Shall be responsible to the Board of Directors. The project manager shall be responsible for guiding and co-ordinating the efforts of all parties engaged in implementing the project, obtaining necessary government approvals on contracts. He is to control the project organisation with the promoters as well as with other agencies and organisations interested in the project. The manager shall have some staff to assist him, especially in checking expenditures to date and determining the present and future cost overrun or under run so that the project manager can take or propose to the Board pertinent corrective measures.

The network shall cover the pre-construction phase of the project indicating major administrative processes, since experience shows that some of them have frequently involved lengthy delays. In other words, it shall include the aggregate activities to be carried out be the principal parties participating in the implementation process.

The project budget shall then be prepared. Part or periodic payments to contractors which might be made at the end of certain intervals (e.g. weekly or monthly) throughout the time horizon of project implementation, shall be made by summing up activity costs per unit of time, which may be a week & month, and computing the cumulative cost at the end of each time interval. For the activities that are in process and are contacted or sub-contracted, the assumption of a linear time cost activity relationship shall be used for the sake of simplicity. In other words, expenditures are uniformly distributed throughout the duration of the activity.

PROJECT SCHEDULE

After an investment decision is taken, the main machinery and long delivery items must be ordered out at the earliest, forming the first major step in implementation of the project. It is foreseen that an engineering consultant will be appointed for carrying out the detailed engineering including basic engineering and procurement assistance to the client. It is also assumed that reputed and experienced contractors with adequate resources viz., men, materials, tools and tackles etc. will be engaged for execution of the construction and erection work. The purchase packages for auxiliaries shall be kept minimum so as to reduce the co-ordination efforts to the minimum. A great deal of co-ordination is required for constructing/erecting the new units. This task is feasible, provided the major activities of the project are co-ordinated and completed in the duration specified to achieve the respective milestones in time.

STRATEGY FOR TIMELY EXECUTION

It is important to deploy a team of experienced personnel for project execution and select the external agencies with due care for rendering the services and supply of equipment for the project. The project activities must be identified, planned and scheduled, and the progress monitored for timely project implementation. All the inputs to the project including financial resources must be identified and their inflow planned and arranged in time.

However this tentative schedule, many activities

will begin in the 6th month after the start of the financing, using different locations and training some partial personal. This will turn possible the strength of the dynamics in the whole process as well as the shorting of the estimated timing.

In short, the following key factors would constitute the broad strategy for timely execution of all activities in a pre-determined manner as per schedule shown in the bar chart as to reach at a basis of regular production.

Early selection of an effective in-house technical team (TASK FORCE) by Government of Mozambique, headed by a Project Manager for planning and executing the project.

i)   Proper choice of external agencies such as consultants for Project Engineering., Machinery suppliers, Construction Agencies etc. keeping in view their reputation/past performance and working experience in their fields.

ii)   Adequate use of computer-based PERT/ CPM techniques for project planning, scheduling and monitoring.

3.6   Description of the initial project Flowchart.

The project implementation phase embraces the period from the decision to start the project to the beginning of the commercial production. It includes a number of stages including negotiations and contracting, project design, construction and start-up.

3.7   Description of the requirement of Land.

For this project the Government of Mozambique will participate with about 25.000 square meters of land. In selecting the land the following criteria should be followed.

1.   The land should be near to the main city of Maputo.
2.   The land should be properly connected by road.
3.   It should also have proximity to many end user clients.
4.   It should also meet various Government policy of
        achieving the social objectives.

2.08 Description of the requirement of Building.

The total requirement of building is as follows.
a. Adm. Building   700 Square Meters

b. Studios –3 (Three)   1.800 Square Meters
c. Other Misc. Building   300 Square Meters

Total Square Meters: approximately 2800 Square Meters.

Cost for construction of about US$750 per square meter (including air conditioning systems, primary electrical cabin etc.).

2.9   Description of some requirements for Plant & Machinery and their basis of selection.

The plant and machinery have been selected giving due consideration to the sophisticated nature of technology required. Detailed discussions were carried out with the foreign suppliers to ensure that the required capacities are practical with minimum capital and operating costs before the machinery were finally selected.

The machinery shall be selected from the most reputed foreign manufacturers of complete range of equipment. Keeping in view the following main factors:

a)   Past performance
b)   Existing machinery in Africa & abroad.
c)   Sales and service facilities in Africa

The main equipment is as follows.

STUDIO

1.   Post-production equipment – non-linear post-production, two ES7 stations.
2.   ENG – 3 DVCAM Camcorders
3.   Régie with 4 DXC-D30PK1 Digital Cameras - for cold light installation.
4.   Flexicart.
5.   Computer graphics suite.
6.   Sound room.
7.   MUST SYSTEM: Automatic Emission Managing System, with graphical information for no active emissions. (This system has been used by several French thematic channels)

Other technical data:
The whole installation is characterised by:
1.   Informational web with UTP level, 5 cables.
2.   Phone system.
3.   Specific furniture.
4.   Parking for about 200 places

Discrimination of Technical Data:

Studio
Cold lighting system.
4 Camera channels composed by:
2 DXC-D30PK Sony Digital Cameras
4 J18 Canon Zoom Lens

4   CA-537P Sony Camera Adapters
4   DXF-50 Sony Studio Viewfinder's
4   CCU-M3P Sony Control Camera Units
4   RM-7P Sony remote CCU controls
4   50M Sony Multicore Camera Sets
4   Canon Focus and Zoom control sets
4   Intercommunication Headphones
4   Special tripods

1 charriot plane and curve

1 Teleprompter with 2 reading systems (monitors for individual speakers) 1 Lighting control table with memory – ARRI
Different microphones with diverse directional characteristics Audio components
Video components
General Intercommunication system

Video Régie

1 Recorder/Player BETACAM equipment
1 video digital mixer with 12 tracks
1 DVE equipment of digital effects for 3 channels
1 TBC Frame Synchroniser
2 Oscilloscope and Vectorscopy for control of cameras
2 Synchroniser Generators with Changeover
1 Intercommunication system with 4 places
1 Audio and Video Matrix with remote controls Video distributors

Audio Régie

1 audio mixer with 24 tracks
2 audio monitors
1 cd player
1 cassette deck
1   DAT
1   Mini-Disc
1   Level Detector for Audio Stereo Stereo Audio Distributors
Audio compressors Audio effects Microphones

Video Post-Production
3 Sony ES-7 on DVCAM/AVID hybrid non linear edition stations 1 Conventional BETACAM Edition Suite 2:1
Multi-track audio (8) on hard disc, Sound Scape type

3 ENG SETS:
Sony DSR-300 PK DVCAM Camcorder Digital Compact Report projectors
Various support materials (Batteries etc.)

3.10   Description of the Water Requirements for the project.

The requirements of water, separately for various matters are given in the following table.

Circulating	: Nil
Make-up	: Nil
Process	: Nil
Drinking	: 3,000 LPD

3.11 Description of the Steam requirement for the project.

A. Steam requirements and steam balance : N.A.

B. Capacity and type of boiler with detailed specifications : N.A.

C. Steam and energy diagram : N.A.

D. Total energy generated / purchased (converted into M. K. : N.A. Cal) theoretical requirement of energy (in M. K. Cal)
at the various consumption stations and expected actual requirement at these stations.

E. If alternate processed are available, comparative energy : N.A.
consumption figures for the various processes. If

the project is energy intensive, possibility of choosing alternate process in order to make the project less energy intensive.

F. Steps proposed to be taken by the company to improve   : N.A.
energy losses efficiency and reduce energy losses (such as power factor improvement, power load management, optimising, illumination waste heat utilisation, etc.)

G. Scope for usage of solar / other renewable sources of   : N.A.
energy.

H. Any other measures contemplated in the direction of : N.A.
energy conservation and management.

3.123 Some information on Compressed air, fuel, etc.

Compressed Air

(a) Requirement   : N.A.
(b) Sources   : N.A.
(c) Arrangements proposed   : N.A.
(d) Cost at site with detailed calculations : N.A.

3.13   Details of the nature of atmospheric, soil

and water pollution likely to be created by the project and the measures proposed for control of pollution. Indicate whether necessary permissions for the disposal of effluent have been obtained.

There shall not be any atmospheric pollution likely to be created by the project as there is no machine which has combustion resulting into air pollution or any chemical process which may release any gases which may result into an air pollution.

However, the use of DG set shall cause a small amount of air pollution. Considering the size of DG set, the pollution is within the permissible limits.

MOZAMBIQUE TELEVISION NETWORK

THE INDUSTRY

CHAPTER - 4

THE INDUSTRY

4.01 Description of the Television Industry in general.

The television industry can be described in the following broad categories. 1. Introduction

2.   Cable Networks

3.   DTH

4.   TV-media

5.   Earnings drivers

6.   Outlook

The detailed discussion in each of the categories is given below.

Introduction

The growing popularity of TV as a communication medium has resulted in the TV media sector undergoing a rapid transformation. From the black and white days of state controlled TV Station, to the highly colourful tunes of Channel V and MTV, the medium has certainly undergone a phenomenal change. Given its popularity, percentage ad spend has also increased proportionately on this medium.

Media pie (%)

	1995	1997
TV	62.5	68.8
Radio	20.9	15
Press	16.6	16.2

Source NRS

Entry of new channels post 1991

All over the world the telecasting has witnessed entry of new channels to cater to the various needs of world audiences. Channels have been launched in English as well as other regional languages. In many countries of the world till 1991, the state owned TV Station ruled the roost, as other players were not allowed to uplink and broadcast. However channels such as CNN, Star TV and BBC, which were offshore companies, could circumvent these regulations and telecast their programs into any country of the world. Cable operators then relayed the same and made it available to the common man through the cable television network.

Like many other countries, the State machinery controlled television. It was used as a propaganda tool for the party in power, with the opposition always at the receiving end. The customer had very little choice. The first steps towards more user choice began during the 1980s, which had to be telecast to a wider audience. TV Station used satellite channels for the telecast and the T.V. network was launched as an international channel.

The sports telecast by Channel 9 in 1985 and the Gulf War in the late eighties all played small but important cameos in educating the international viewer. With liberalisation in 1992 and crumbling tariff barriers televisions (read as colour TVs) became more easily available. The media revolution had started.

Major satellite channels avidly watched by viewers are Star TV, Sony TV, Home TV, BBC & CNN. There are other regional language channels which are major players in their respective territories.

Most of the channels that could not attain popularity rapidly suffered, since their advertisement earnings were not sustainable. The first round of the media wars is over. Management changes, i.e.

original promoters selling out to new management with deeper pockets, has become the order of the day. Alliances like the famous ESPN Star Sports arrangement also made headlines. Given the global trends of mergers and acquisitions, further consolidation is likely. Alliances and mergers make sense when the partners complement each other, like BBC and Discovery launched Animal Planet, CNBC and ABNI came together to launch a business channel called CNBC Asia.

Cable Networks

Antennas set up by either the end user or the cable operator receives the signals transmitted by the satellite. Local cable operators lay their own cables, set up control rooms, which can telecast 40 or more channels over a limited area. They charge the household a one- time connection charge of about US$.10 per point and a recurring monthly charge ranging between US$.1 to US$.5.

Initially, this was done in a very unorganised manner. The business required local knowledge and contacts, so every locality had its own cable operator. Collection was critical for the cable operator. For the end user, quality of telecast and a complete lack of standards became an issue. This lead to a shakeout and the formation of cable companies with money power which in turn tied-up with the local and small cable operators. Cable companies charge about US$1 per month to the local cable operators and support them with training and other infrastructure inputs. The business is immensely capital intensive and takes a long time to break even.

In many countries the operations of cable operators are regulated under the Cable TV Act which ensures that pornographic materials and other materials which are against culture and values or are detrimental to national interests do not get telecast. Recently this act has been amended to include foreign channels also.

Direct To Home

DTH is a new technology that circumvents the cable operators by directly delivering a bundle of channels to the end user. DTH involves transmission of encoded audio/ video signals (Ku band) via satellite. The end user needs an antenna to receive the signals and a decoder (set top box) to unscramble the encrypted signals. DTH services elsewhere in the world are Echostar and DirecTV (USA) and BskyB (Europe). Rupert Murdoch of Star TV fame owns BskyB.

The size of the antenna in DTH will be 1.5-2 ft in diameter, making it easy to install and transport. In conventional cable, since signals are in C band, an

8ft- diameter antenna is needed. The basic difference in the business model is the hardware costs in DTH. In a cable system, the user pays a one time connect fee and monthly rentals, while in DTH he has to invest in hardware.

The antenna will cost about US$200-300 and decoder will cost about US$200. The African viewer might be reluctant to incur such heavy installation costs. Quality of telecast in DTH is superior to Cable TV and viewer can receive up to 200 channels.

DTH will result in restructuring of the cable television industry. It will become imperative to have cash reserves to withstand the technology threat. Up gradation to fibre optic backbone will become necessary. A fibre optic network will cost about US$0.5mn per km as compared to US$0.1mn per km for coaxial cable. The stage is now ripe for consolidation.

TV Media features

In the Broadcasting business, it is only the industry leader who makes sizeable profits. The business is a game of asymmetrical payoffs. For instance, the top 5 channels account for 90% of ad spend.

Urbanisation and TV penetration is related. This may be due to the popularity of cable television that has resulted in increased colour TV sales. Rural penetration is low, although growing at a fast pace, because of dearth of specific program content to cater to that segment.

Liberalisation has resulted in the world viewer becoming more aware and conscious. This has resulted in the customer having more choice with the entry of a number of companies in different segments. Competition has resulted in companies increasing their marketing spend significantly.

Popularity of TV media is becoming higher. Increasing TV penetration leads to a reallocation of advertisement budgets with higher allocation for television at the cost of other medium.

TV channel operators use different business models to generate revenues. The critical component of any channel is the quality and type of programs they telecast. This determines their popularity, which in turn determines amount of advertisement revenues they can generate. They can do any one of the following:

Buy programming rights of program software from outside and collect advertisement revenue on their own. This model is followed by several TV companies, wherein they have a separate company in their fold, which develops all the content. The advantage is that re runs of serials/ programs become very profitable.

Selling time space to the producers for a fixed charge. Producers in turn are free to book advertisements at their own rates (there is an understanding on the time allocated for advertisement) and collect revenue. This is the basic model for many TV companies in which they sell prime time slots. The rights continue to be vested with the producer.

Earnings drives

The key factors that drive sector revenues are

Television penetration: Since the medium is television, increased television penetration will imply higher viewer ship. This will translate into higher advertisement spend allocation. This will also imply higher software production and demand for new programs.

Competition from other satellite channels would have an adverse impact on advertisement revenues, as advertisers have more choice in allocating ad budgets.

Government policies can have a big impact on the fortunes of the entire industry. When the DTH bill is passed in any country then, it will trigger a restructuring of the cable business.

Launching new channels targeted at specific segments, like regional channels within any country
other areas having large pockets of ethnic population would lead to revenue growth. This will entail
significant initial outlays.

Depreciation of the local currency would increase
revenues as most of the program/ software companies export the programs overseas and payments
are dollar denominated.

Advertisement revenue

As mentioned earlier, this is the primary source of
income for TV channel operators. This revenue is
directly co-related with the reach and viewer ship
of a channel. Any channel's popularity depends on
good quality programs, which is the software content. The business requires enormous initial investment in programs and revenues follow only with
a time lag after the channel receives a minimum
viewer acceptance.

Outlook

The sector has latent potential for growth on back
of the exponential growth of cable TVs during the

last 5 years. Television penetration in Africa is extremely low as compared to other developing countries like Malaysia, Pakistan, etc. in Asia. The number of channels has increased, implying higher demand for software programs.

Advertisement revenues, which are the barometer of channel popularity, will get dispersed over several competing channels. A shakeout is likely in both the channel and cable TV sectors. The biggest beneficiaries will be the content providers or the software houses. They will control the intellectual rights to the key element driving any channel's popularity.

Direct-to-Home, Digital Terrestrial Transmission and Conditional Access Cable Delivery have emerged as new delivery mechanisms. Breakthrough in technology would help open up avenues for these channels.

MOZAMBIQUE TELEVISION NETWORK

MARKETING

CHAPTER - 7

MARKETING

7.1   Description of the commercial viability of the project with regards to revenue generations.

Estimations shows that, after about 5 years of regular working, the CPLP Cultural Television Channel will be able to start exporting a great quantity of services and videos to Africa, Europe and South America. (See financial projections)

The economic feasibility of the project will occur with the commercialisation of several products, like:

1.   The sale and production of cultural programmes to various institutions like:
-    BBC
-    TV Culture Brazil
-    RTP
-    Discovery
Etc.
All over the world commercial networks for tel-

evision cultural programmes have been created. Several television channels in different countries have regularly acquired programmes by producers spread out all over the world. Not only, several international organisations, like the World Bank and the United Nations, including FAO and Unesco, among many others, need programmes for public education, like programmes oriented to alert and to educate people concerning endemic and epidemic diseases. AIDS, malaria, tuberculosis are a very few examples.

These world institutions have made a great effort to develop television programmes with humanitarian objectives. But, in Africa – undoubtedly the continent most needed of such programmes – there is no television station, in present times, with capability to make face to such a need.

Therefore, the television teams that had been responsible for such programmes are, practically in its totality, placed in countries of the called First World, strange to the local population's concrete reality.

Programmes focusing new agricultural techniques or even oriented to agricultural, commercial and industrial education are essential elements of such a repressed demand.

The price for each institutional campaign, with an

average of 10 films produced per campaign, is of about USD.15.000$ and the capacity of the CPLP Cultural Channel will be as follows:

First Group* - from 2 to 4 campaigns per month.
Second Group – from 1 campaign in two months to 1 campaign per
month

This represents:
First Group   a potential annual income of about US$.540.000
Second Group a potential annual income of about US$.90.000

*First Group: Angola, Mozambique, and Mozambique. Second Group: Guinea-Bissau, Sao Tome e Principe, East Timor (Lorosae). Reference Group: Brazil, Portugal.

Obs. In the Second Group, for the first six years of operation it is estimated a difference of potential annual income between the Angola unit and Mozambique's and Mozambique's respective units, from US$.540.000 to about US$.378.000.

2. The sale and production of no cultural programmes to other television channels, like:
-       series (novels)
-       talk-shows Etc.

Many countries of the region need to produce television programmes, but do not have capability to do it. Thus, they seem themselves obliged to search expensive productions in Europe. With only one programme sold-three daily hours-the channels of the First Group* would receive in incomes the equivalent of about two times of the whole investment. The Second Group countries will not have conditions at the beginning to product cultural programmes to other television channels.

*First Group: Angola, Mozambique, and Cape Verde. Second Grope: Guinea-Bissau, Sao Tome e Principe, East Timor (Lorosae). Reference Group: Brazil, Portugal.

The price for this type of television programme is:

Auditory programs   US$.12.500 per programme
Talk-shows   US$.4.000 per programme
Interviews   US$.4.000 per programme

The capacity of production, of these programmes by the countries of the First Group will be:

In the FIRST PHASE:
Talk-shows   2 per month

Interviews   4 per month
In the FINAL PHASE:
Auditory programmes   6 per month
Talk-shows   20 per month
Interviews   20 per month

Thus, the potential income will be:

FIRST PHASE
Talk-shows   US$.96.000 per year
Interviews   US$.192.000 per year

FINAL PHASE
Auditory programmes   US$.900.000 per year*
Talk-shows   US$.960.000 per year*
Interviews   US$.960.000 per year*

* the same difference as showed above.

The price of the novels is much higher. For each complete novel, with about 120 chapters, the price is about US$.2.500.000 and the capacity of the First Group of the CPLP Cultural Channel (in the final phase) will be of one novel per year.

3. There is the possibility of commercial use of the transmission time beyond the six hours reserved to culture and education.

The conventional day in television is of 18 hours,

of which only 6 hours would be no commercial. We would have, therefore, 12 hours of no cultural transmission, which should be freely commercialised.

Even the period of six hours- divided into two sections of three hours, the first one dedicated to culture and second section of three hours to education-has a great potential for sponsoring.

Each commercial hour of transmission can include up to 12 minutes of advertising. The price for each 30 seconds of advertising is:

Noble time   from 6PM to 11PM   a b o u t US$.400 each 30"
Normal time   rest   about US$.200 each 30"

Concentrating the commercial programming in the noble time. With 4 hours of transmission in this period, we would have:

Educational television: from 7AM to 9AM
From 3PM to 4PM
Cultural television: from 4PM to 7PM
Commercial television: from 9AM to 3PM
From 7PM to 1AM

Thus, the commercial period would comprehend 6 hours in normal time, 4 hours in noble time and 2 hours in normal time again.

Begin 12 minutes per hour for advertising, the commercial period could have:

Noble time   48 minutes   24   films   US$.9.600 per day

Normal time   96 minutes   48   films   US$.9.600 per day

Total of the potential income............   US$.19.200 per day

> or US$.576.000 per month
> or US$.6.912.000 per year

It is believed that in the first phase of the project (after one year), the CPLP Cultural Television Channels of the First Group will be able to start with an income from commercial advertisement of about US$.1.000.000 per year-value which is predicted to increase in the follow months.

4. The rent of the studios to television teams of other countries can be another source of incomes. Many producers who cover events in Africa need to move many times to their countries of origin during the video works, because there is not technical support in Africa. The same phenomenon happens with the cinematography and the journalism productions. It is not difficult to imagine, for example, the serious problems journalism teams have had, for example, with essential components like batteries, lighting, electronic components etc., which only can be easily find in Europe.

The period of renting of a television studio is of 12 hours. Each period has price of about US.4.000. After the first year, the studio should be rented for 12 hours per each period of 3 days. So, the rent per month, in this period (First Phase), will be able to generate incomes of about US$.120.000 per month in the First Group.

5. Support services to other television networks in all areas, including novels, mini series, docdramas series etc.

Many countries, principally in Africa, do not have technical conditions to develop this kind of programmes, but they have a strong internal repressed demand in this sense.

Each team, abroad, has a price of US$.1.000 per day (two people), after the costs of dislocation, hotel and meals. In the first months (first phase) it will be organised one team for works in other countries. In the final phase it is predicted to have up to 5 teams with such function. The capacity of each team is of about 20 days per month. Therefore, the support to other television networks-when took in its full potentiality-will be able to represent up to 20 days per month in the first phase. This represents a potential income of about US$.240.000 per year, and 100 days per month of works in the final phase, signifying about US$.1.200.000 per year of incomes, always referring to the First Group countries.

6. Colloquies, seminars, videoconferences and meetings of different natures.

The building of the CPLP Cultural Television Network will have a medium size auditorium, (First Group), with all conditions to receive seminars, colloquies and meetings of the most diverse nature. Such seminars, videoconferences etc. are important not only for the increase of the incomes, but also to attract specialists of the most different areas, being an important element for the development of all region as well as for the diffusion of the local, regional and continental cultural values.

The meetings, seminars and colloquies attract, in average, about 300 people per event. The price-excepting meals, hotels and transportation-per each participant is of about US$.50 per day. It will be a capacity for up to four events of this type per month, what could represent an income of about US$.60.000 per month or US$.720.000 per year (final phase-after 5 to 7 years).

The project should also turn possible.
-        classes of professional formation in diverse disciplines
-        support for a multimedia high technology centre (First Group)
-        support for a multimedia high cultural centre (Final Group)

MOZAMBIQUE TELEVISION NETWORK

PROFITABILITY & CASH FLOW

CHAPTER - 8

PROFITABILITY AND CASH FLOW

8.1   Estimation of cost of production and working results for the first five years of operation.

The estimated cost of production of working results for the first five years of operation are given in the chapter "Financial Projection".

8.2   Cash flow statement for the company as a whole, for five operating years of the project based on the estimates of working results.

A detailed cash flow statement for the company as a whole for five operating years is given in the chapter of "Financial Projection".

8.3   Projected balance sheet for five operating years for the company as a whole.

The balance sheet for five operation years for the company as a whole is given in the chapter of "Financial Projection".

MOZAMBIQUE TELEVISION NETWORK

ASSUMPTIONS

CHAPTER - 9

ASSUMPTIONS

1.   The Mozambique Television Network shall generate the income from the following sources.

a)   Sale of cultural programmes.
b)   Sale of commercial programmes.
c)   Sale of advertisement time during transmission of TV programmes

d)   Hiring of studios.
e)   Supply of technical services
f)   Hiring of conference hall

The detailed assumptions for each of the above mentioned activities are as follows:

a)   Sale of cultural programs:

I.   It is assumed that four cultural programs shall be produced per month in the first year, which can be sold to other countries. This
figure will increase to 5 programs per month in the second year, six programs per month in the third year and so on.
II.   It is assumed that the selling price shall be US$.105 per program.

b)   Sale of commercial programs:

I.   The commercial programs consist of talk shows, interviews and auditory programs.
II.   It is assumed that two talk shows shall be produced per month in the first year, three talk shows shall be produced in the third year, four talk shows per month shall be produced in the third year and
so on.
III.   It is assumed that the selling price of one talk show program shall
be US$.2800.

IV.   It is assumed that four interviews shall be produced per month in the first year, five interviews shall be produced in the third year, six interviews per month shall be produced in the third year and so on.

V.   It is assumed that the selling price of one interviews program shall be US$.2800.

VI.   It is assumed that no auditory programs shall be produced in the first and the second year. Only when the people have two years of experience then in the third year one auditory program shall be produced per month. In the fourth year two programs shall be produced per month and so on.

VII.   It is assumed that the selling price of one auditory program shall be UD$.8750.

c) Sale of advertisement time during the transmission of T.V. programs:

I.   There shall be a total transmission of 8 hours per day in the first year. It will go on increasing by two hours in the second and subsequent years.

II.   Out of which advertisement shall be available for programs with transmission period of four hours in the first year. The programs in which advertisements could be available shall increase by two hours per year in the second and subsequent years.

III.   In one hour T.V. transmission, 12 minutes of advertisement shall be allowed.

IV.   Out of total T.V. time 33 % time shall be considered as prime time and remaining 67% shall be considered as non prime time, in the first, second and third years. From the fourth year onwards the prime-time advertisements shall remain constant.

V.   The advertisement rates shall be US$ 560 for one minute of prime time advertisement and US$ 245 for one minute of non prime time advertisement.

d)   Hiring of studios:

I.   It is assumed that the studios shall be taken on hire for a shift of 12 hours, two times in a month in the first year. It will increase to three times in a month in the second year.

II.   The rent per day shall be US$2800.

e)   Supply of technical services:-

I.   It is assumed that one team, consisting of two technically qualified people shall be available in the first year. In the second year two such teams shall be available. In the third year three such teams shall be available and so on.

II.   It is assumed that in the first year the team shall be hired for 10 days in a month. In the second year the teams shall be hired for 12 days in a month. In the third year the teams shall be hired for 14 days in a month and so on.

III.   The rate of hire shall be US$.700 per team per day.

f)   Hiring of conference hall :-

I.   It is assumed that the conference hall shall be taken on hire for two times in a month in the first year, three times in a month in the
second year and four times in a month in the third year and so on.
II.   The hire charges shall be US$.350 per conference.

2.   The unit is planning transmission of three hours of educational programs, three hours of cultural programs and two hours of commercial programs per day in the first year. Thus, there shall be a total transmission of eight hours of programs per day in the first year. The total transmission hours shall increase by two hours per day in the second and subsequent years. Thus, in the second year there shall be total transmission of 10 hours per day and in the third year there shall be a total transmission of 12 hours per day. The cost of software production is assumed at US$.1750 per hour.

3.   The cost of miscellaneous consumable items is assumed at about US$.25200 in the first year. This will increase by 10% in the second and subsequent years.

4.   The cost of stores and spares consumed is assumed at US$.29400 in the first year. It will

increase by 10% in the second and subsequent years.

5.   The cost of repairs and maintenance is assumed at 0.1% of the total cost of plant and machinery in the first year. This cost will increase up to 0,3% in the fifth year.

6.   It is assumed that the unit shall use 1.000 units of power every day or 365.000 units of power in the first year. The consumption of power shall increase by 10% in the second and subsequent years. The cost of power is assumed at 3.5 cents for one unit.

7.   It is assumed that the unit shall consume 700 litters of diesel per month. The cost is assumed at 33 cents for one litter. It will increase by 10% in second and subsequent years.

8.   The unit shall have a technical staff of 35 people and administrative staff of 15 people.

9.   Depreciation is calculated on reducing balance method. The rate of depreciation is taken at 10% for building, 25% for plant and machinery, 10% for furniture and fixtures and 20% on vehicles.

10.   Advertisement expenditure is assumed at 0.2% of the sales.

11.   It is assumed that the stock of raw material and stores shall be 30 days of consumption, receivables shall be 45 days sales and creditors shall be 15 days purchases.

12.   The term loan shall be for a total period of 25 years. It shall have a moratorium of five years. It shall be repayable in 20 equal yearly instalments.

PROJECT REPORT

OF

SAO TOME CULTURAL TELEVISION NET-WORK

AT

SAO TOME E PRINCIPE

AFRICA

PROMOTERS

REPUBLIC OF SAO TOME E PRINCIPE

AFRICA

SAO TOME TELEVISION NETWORK

INTRODUCTION

CHAPTER - 1

INTRODUCTION

1.1   Description of the organisation through which this project is being taken up.

The CPLP is an international organisation, created in 17 of July of 1996, with Headquarters in Lisbon and consisting of the following State Members: Angola, Brazil, Cape-Verde, Guinea-Bissau, Mozambique, Portugal, Sao Tome e Principe and East Timor (Timor Lorosae).

The CPLP has as its objective the politic-diplomatic relationships between its Members, mainly referring to the United Nations and the World Bank, the relationship EU/MERCOSUL and the accomplishment of the Europe-Africa Summit.

The CPLP is also dedicated to the co-operation, particularly in the economic, social, cultural, legal and technical-scientific matters; as well as to the projects of promotion and broadcasting of the Por-

tuguese Language, nominated to the improvement of the International Institute of the Portuguese Language and the creation of a Bibliographical Fund.

All its decisions are taken by consensus.

2.2   Description of the main reasons for setting up an African Cultural Television Network.

Africa is a continent submerged into dramatic problems. All these humanitarian problems have in the education and culture their most direct and objective root.

The African countries speaking Portuguese are among the African countries with the lowest level of development.

The whole situation is so dramatic that the implementation of the CPLP Cultural Television Network is more than urgent.

It is hoped that with this Television Network the people shall have better understanding of various diseases and other problems of life. e.g. AIDS is prevalent disease in Africa. But as the people are illiterate the information about the disease cannot be spread by any other method. In such cases the Television Network becomes a very effective audio-visual means of communication.

The project also intends to be a base of a massive production of health campaigns, specially to reinforce the combat against the AIDS, the infect-contagious diseases- endemic or epidemic-through a process of global education of the local and regional populations.

The CPLP Cultural Television Network is a project for peace and its objective is not to compete with local television channels already installed, but to interchange with them, promoting a future global web dedicated to knowledge and development.

1.03 Description of the main objects of the project.

The major objective of the project is to assist the CPLP countries to develop the first African television network channel oriented to cultural and educational themes with the follow orientation:

- diffusion of the Portuguese language
- education in all areas
- reinforcement of the local and regional cultures
- formation of technical personal on communication and tele-education
- formation and information for a better participation of the CPLP nations in the process of global development
- to support the implantation of the Institute

of Portuguese Language

The development of such a project will create an irradiation pole from the countries with the project implementation to the whole African continent, constituting an African network of research and culture, discovering and valorising different cultural aspects and, consequently, reinforcing identity factors.

With the stronger cultural identity, it is expected to have lower levels of violence in social behaviour as well as a relevant increase of the educational performance in all its sectors.

Systematic educational objectives are complementary to the general orientation of this project, which is cultural in its fundament.
A very large positive impact is also to be expected with respect to several sectors of the economy, greatly improving the overall quality of services.

1.4   Description of the mission status of the project.

The mission of the project is:

-       to rescue lost values and cultural information
-       to reinforce historical elements

- to rescue old artworks
- to give a new value to new artworks
- to communicate practical and objective information on health
- to communicate practical and objective information on agriculture
- to communicate practical and objective information on economy

1.05 Description of the philosophy of the project.

The CPLP Cultural Television Network will have in its philosophy, as a support to its mission, the follow fundaments:

- to work with local people
- be focused on quality
- intercommunication

Three words can define the CPLP Cultural Television Network philosophy:
People, quality and intercommunication

Following these three key words of its philosophy, a single sentence also defines its mission: "A Continent of Culture".

Two main sectors characterise its global structure:
1. Cultural Sector
1.1.   To develop programmes on:
-       Music

- Architecture (also vernacular)
- Archaeology
- Literature and poetry
- History
- Tourism
- Dance
- Plastic arts
- Astronomy
- Science in general
- gastronomy
- health
- anthropology
- local cultures
- others

1.2   The programmes can be of all types, the most important is to be attentive to communication. All programmes must reach the highest possible number of spectators.

2. Educational sector.
2.1   Systematic education in short practical courses.
2.2   Educational series with practical and objective information on:
- health care
- family
- agriculture
- managing
- services
- civil construction

As the CPLP Cultural Television Network must not be supported by the respective governments in the future, it will have a daily time of free commercial programmes up to 12 hours. This is a good solution to pay its costs and support its development.

1.06 Description the magnitude of the project.

The project proposes a Television transmission time of eight hours per day in the first year. It will increase by two hours per day in the second and subsequent years.

The project also envisages training of local people in all the related areas of Television. It is proposed to develop a local pool of talent which can be utilised not only in this particular country but also in other countries of Africa.

The total cost of the project of the Sao Tome Cultural Television Network is US$.13.200.000. Out of this it is proposed to obtain a soft loan of US$.11.000.000. The remaining amount of US$.2.200.000 shall be contributed by the Government of Sao Tome in terrain for the installation of the respective building, infrastructures and logistic support.

For the purpose of the Sao Tome Cultural Television Network a company shall be formed with the participation of the Government. The Contribution of the Government in the form of share capital shall be US$.4.500.000.

The term loan shall be on soft terms. The total period of the loan shall be 23 years. Out of which five years shall be a period of moratorium. The entire loan shall be repaid in the remaining 17 years, in 17 equal yearly instalments.

The Government of Sao Tome will give the sovereign guarantee for the referred loan.

SAO TOME TELEVISION NETWORK

THE COUNTRY

CHAPTER - 2

THE COUNTRY

2.01. Description the level of development of Sao Tome.

This small poor island economy has become increasingly dependent on cocoa since independence over 20 years ago. However, cocoa production has substantially declined because of drought and mismanagement.

The resulting shortage of cocoa for export has created a persistent balance-of-payments problem. Sao Tome has to import all fuels, most manufactured goods, consumer goods, and a significant amount of food. Over the years, it has been unable to service its external debt and has had to depend on aid and debt rescheduling. Considerable potential exists for development of a tourist industry, and the government has taken steps to expand facilities in recent years. The government also has attempted to reduce price controls and subsidies, but economic growth has remained sluggish. Sao Tome is also optimistic that significant petroleum discoveries are forthcoming in its territorial waters in the oil-rich waters of the Gulf of Guinea.

OFFICIAL NAME: Democratic Republic of Sao Tome and Principe
CAPITAL: Sao Tome
SYSTEM OF GOVERNMENT: Multiparty Republic
AREA: 963 Sq Km (372 Sq Mi)
ESTIMATED 2000 POPULATION: 148,700

LOCATION & GEOGRAPHY: Sao Tome and Principe are islands located in the Gulf of Guinea off the coast of West Africa. Both islands are of volcanic origin and are characterised by many craters and lava flows. Sao Tome has ten peaks over 1,067 metres (3,500 feet) while Principe has a larger and flatter area than Sao Tome. Streams radiate throughout the forest clad mountains to the sea on both islands. Major Cities (pop. est.); Sao Tome 43,400 (1991). Land Use; pastures 1%, agricultural-cultivated 39%, forested and other 60% (1993).

CLIMATE: Sao Tome and Principe has a tropical climate characterised by hot and humid conditions that are influenced and modified by the cold Benguela current as well as by altitude. The dry season is from June to September and the wet season from October to May. Average annual precipitation varies from 500 mm (197 inches) on the south-west mountain slopes to 1,000 mm (394 inches) on the northern lowlands. Average temperature ranges in Sao Tome are from 21 degrees Celsius (70 degrees Fahrenheit) to 31 degrees Celsius (88 degrees Fahrenheit) with the temperature of the interior's higher altitude around 20 degrees Celsius (68 degrees Fahrenheit) .

PEOPLE: The principal ethnic majority is formed by Africans. Other ethnic minorities include Angolares who are descendants of Angolan slaves and the Portuguese.

DEMOGRAPHIC/VITAL STATISTICS: Density; 123 persons per sq km (319 persons per sq mi) (1991). Urban-Rural; 40.5% urban, 59.5% rural (1988). Sex Distribution; 49.4% male, 50.6% female (1991). Life Expectancy at Birth; 64.0 years male, 67.0 years female (1990). Age Breakdown; 38% under 15, 22% 15 to 29, 17% 30 to 44, 12% 45 to 59, 8% 60 and over, 3% unspecified (1989). Birth Rate; 38.0 per 1,000 (1992). Death Rate; 8.0 per 1,000 (1992). Increase Rate; 30.0 per 1,000 (1992). Infant Mortality Rate; 58.0 per 1,000 live births (1992).

RELIGIONS: Mostly Christians with 84% of the population Roman Catholic while the remainder follow local native tribal beliefs.

LANGUAGES: The official language is Portuguese

which is spoken in a heavy Creole dialect.

EDUCATION: Aged 25 or over and having attained: no formal schooling 56.6%, incomplete primary 18.0%, primary 19.2%, incomplete secondary 4.6%, secondary 1.3%, higher 0.3% (1981). Literacy; literate population aged 15 or over 28,114 or 54.2% (1981).

CURRENCY: The official currency is the Dobra (Db) divided into 100 Centimos.

ECONOMY: Gross National Product; USD $41,000,000 (1993). Public Debt; USD $225,800,000 (1993). Imports; USD $30,400,000 (1994). Exports; USD $6,500,000 (1994). Tourism Receipts; USD $1,000,000 (1990). Balance of Trade; USD -$23,900,000 (1994). Economically Active Population; 49,216 or 41.0% of total population (1991). Unemployed; 22.0% (1994).

MAIN TRADING PARTNERS: Its main trading partners are Portugal, Angola, the Netherlands, the USA, Germany and the UK.

MAIN PRIMARY PRODUCTS: Bananas, Cocoa, Coconuts, Coffee, Fish, Palm

Kernels, Timber.

MAJOR INDUSTRIES: Agriculture, Food Processing, Light Construction,

Timber Products.

MAIN EXPORTS: Cocoa, Coffee, Copra, Palm Kernels and Nuts.

TRANSPORT: Railroads; nil. Roads; length 380 km (236 mi) (1988). Vehicles; cars 1,774 (1975), trucks and buses 265 (1975). Merchant Marine; vessels 3 (1990), dead weight tonnage 1,172 (1990). Air Transport; passenger-km 6,100,000 (3,790,000 passenger-mi) (1985), cargo ton-km 100,000 (68,490 short ton-mi) (1985).

COMMUNICATIONS: Weekly Newspapers; total of 2 with a total circulation of N/A . Radio; receivers 31,000 (1994). Television; 21,000 (1994). Telephones; units 2,400 (1993).

MILITARY: N/A. Military expenditure accounts for 1.6% (1980) of the Gross National Product (GNP).

SAO TOME TELEVISION NETWORK

THE PROJECT

CHAPTER - 3

THE PROJECT

3.1   Description of the initial steps to be taken for the implementation of the project.

The first objective of the CPLP Cultural Television Network is to install its structures in the following countries:

-        Sao Tome e Principe
-        Cape Verde
-        Mozambique
-        Guinea-Bissau
-        Angola
-        Timor

Thus, these six countries will structure the basic network of the whole project.

Angola, Mozambique and Cape Verde will form a FIRST GROUP; Sao Tome e Principe, Guinea-Bissau and East Timor will form a SECOND GROUP; and, finally, Brazil and Portugal will form a called REFER-ENCE GROUP.

However, the project will start at the same time in all countries. A more precise description of the meaning of each group is made below.

The basic physical components of the project are:

1.   The creation of companies, in each of the States of the CPLP, oriented to the objectives above referred.

2.   The State of each country – who will be the responsible and guarantee of the payment of the financing – will also have a participation in the companies.

3.   All strategies will be oriented to lowest costs in long term.

4.   The whole project will be developed by phases.

5.   The timetable of transmission will be defined case by case.

6.   The project will include:

6.1.   a complete plan for an equipment network.

6.2.   a complete plan for a software network.

6.3.   all procurements

6.4.   equipment transportation and installation

6.5.   all configurations

6.6.   creation of the infographic system

6.7.   technical formation of the personal

6.8. architecture and engineer projects for the building

6.9. construction of the building organised in phases

1.2 Description of the procedure by which the necessary programmes will be produced for transmission on the channel.

The television programs are known as software in this field. The unit is planning transmission of three hours of educational programs, three hours of cultural programs and two hours of commercial programs per day in the first year. Thus, there shall be a total transmission of eight hours of programs per day in the first year. The total transmission hours shall increase by two hours per day in the second and subsequent years. Thus, in the second year there shall be total transmission of 10 hours per day and in the third year there shall be a total transmission of 12 hours per day. The cost of software production is assumed at US$.2.500 per hour. The amount of US$.2.500 per hour is taken based on cost of production of such programmes in Portugal.

To contract a similar system from other television channel, only for one country, for a transmission period of 6 hours daily, the cost in, market (eg. Portugal*) is of about US$.500.000 per month. This cost represents a total of about US$.6.000.000 per year. In 16 years, excluding the first year of implementation, the costs for contract the six daily hours of transmission from other television channel, in present market price, is of about US$.100.000.000. The six countries would have an estimated cost of about US$.600.000.000 in a 16 years period of time.

These costs are based on the prices presently used in Portugal, as for example the programme Hora Viva – Seguranca em Directo, transmitted every day, excluding weekends, from 7 to 10 AM.

Another example is the channel Canal Noticias Lisboa CNL. Only in its first year of installation, it was expended about US$.10.000.000, and the equipment as well as the building were not property of the channel, but rented.

In fact, however the project will be implemented by phases, after 5 to 10 years of regular work, it is predicted to have reached a similar value in incomes as the examples above.

*value took in the Portuguese market, below the average of the European prices in about 30%.

The first ten months of works will be oriented to:
-         design of the equipment and software struc-

tures
- start the first productions
- elaboration of the projects of architecture and engineering
- procurement and acquisition of all equipment
- construction and supervision of the works
- installation and configuration of the equipment

So, the first year must be used to the implantation of the project. However, as to accelerate the chronogram, some operations should be started before this timing, temporarily installed in a different building.

The first phase will be oriented to:
- video productions (external and internal)
- edition
- elaboration of cost programmes

3.3   Description of the requirements of Power, Fuel and Water. POWER
It is assumed that the unit shall use one thousand units of power of everyday. Thus, the unit shall use 365.000 units of power in a year. The costs of power is assumed at five cents for one unit. The consumption of power shall increase by 10% every year.

FUEL

The unit shall use D.G. set for generation of electricity, when there is an interruption in the supply. The hours of disruption in the normal supply can not be predicted accurately. Hence, the hours for which the D.G. sets shall be used also can not be predicted accurately. However, it is assumed that the unit shall utilise one thousand litters of diesel per month. The cost is assumed that 33 cents for one litter. Hence, the total cost for one thousand litters shall be US$.330. The consumption of diesel shall increase by 10% every year.

WATER

The unit does not need water for fixed commercial operation. However, it shall have a strength of 50 people as employees. Water shall be required for drinking and sanitation purposes. It is assumed that 3000 litters of water shall be required everyday. The cost of obtaining this water is assumed at US$.10 everyday. The consumption of water shall increase by 10% every year.

3.4   Description of the manpower requirement of the project and description of the steps taken by the company to train the manpower.

The total staff strength shall be 17 people in the first year.

The system of modern TV Station is designed to achieve sustained operating efficiency and transmission. This, of course, entails a certain degree of sophistication in the production and transmission. However, considering the socio-economic situation in Sao Tome, a reasonable balance has to be struck to obtain optimum performance and at the same time create gainful employment. While working out the manpower requirement for this project to be kept on direct rolls of the company totalling 50 staff and operators, the above consideration have been kept in mind.

Manpower requirement

The direct manpower required for the proposed Unit is about US$.1.800.000 per year. The manpower requirement as indicated in this chapter has been planned keeping in view the following guidelines:

Effective co-ordination among the various departments. Judicious distribution of responsibilities. Capacity utilisation of the TV Station with optimum manpower.

Details of manpower requirement for the TV Station is given in the financial section.

In line with the prevailing practice, the Security guards, office peons and unskilled labour etc., are normally employed on contractual/casual basis, and their cost has to be included in the Factory over-heads. However in this report provision has been made to employ the above staff and their salaried have been included.

The various departments proposed shall be under the direct responsibility of the Station Director. He shall be assisted by Deputy Directors, who will look after the complete TV Station and its various day to day activities and Marketing. They shall be respon-sible for achieving the envisaged targets and sales forecasts. They shall be assisted by a team of Man-agers from production, marketing and accounts.

To run this project a labour force which shall usu-ally be composed of unskilled and skilled workers shall be employed. The first are those who do not undergo any king of specific training or education, while the latter have to do so in order to master their jobs.

When evaluating an investment project from an em-ployment point of view, its impact on both unskilled and skilled labour has been taken into account. Not only direct employment, but also indirect employment has been considered. Direct employment refers to the new employment opportunities created within the project; indirect employment concerns job opportunities created in other projects linked with the project which is being formulated.

The implementation of large and sophisticated projects generally contributes to the development of local skills and capabilities in a country. Furthermore, they help to change traditional values, attitudes and the behaviour of the society, to build up an enterprising spirit among the people, to develop a desire for changing and improving the existing conditions of life, to introduce better work discipline and thus to change the very pattern and basis of economic development. The TV industry is already well established in Africa. Location of TV industry activities in Africa, has certain favourable factors and advantages, in setting up this type of industry, as availability of acceptable level of education among the supervisory staff adequate technical and managerial skills developed over a long period, and availability of cheap labour are assured.

The organisation structure will vary from TV Station to TV Station in the industry and as such the pattern proposed herein can be considered only as suggestive and provisional.

A well-knit organisation structure headed by a Station Director and Manager, with the supporting staff will be developed progressively during project implementation. Soon after the plant becomes operative, a good number of project staff will be absorbed in the organisation. Certain additional staff also get added to ensure smooth and efficient management of the operating unit.

The Manager Production will have a degree in communication and will be accountable and answerable to the Station Director for all TV operations including planning, production, material management, TV utilities, quality control, production cost and budgetary control, TV safety, discipline and layout relations. His main function should be to ensure achievement for quality and quantity targets of production at reasonable cost and should constantly strive to improve TV performance. In the discharge of his multifarious duties and responsibilities he will be assisted by supervisors and adequate staff for day to day activities.

His duties are to ensure that targets are maintained through effective utilisation of 3 M's viz: Machinery, Men and Materials. These targets should be translated in terms of targets for individuals under him such as supervisors, skilled workers etc. He has to ensure that the equipment at his disposal gets prompt attention on breakdown is adhered to strictly. He has to further ensure that there is a strict quality control exercised over raw materials also. He will be assisted by Deputy Manager etc. who and will report to the General Manager.

GENERAL MANAGER (Comm.) He will report to the Station Director. His main duties will be prepare sales forecasts and budgets, study and monitor the export market for the company's products and advice on ways and means of increasing sales. He will also ensure that consumer complains are solved in an appropriate manner. He will also ensure that the right materials are available for production at the right time.

He will be responsible for all aspects connected with the export procedures and keep the management updated on all matters relating to the Govt. policies on polymers and Exports and above all the world market.

COMPANY SECRETARY & FINANCIAL CONTROLLER. He will report to the Station Director and will be responsible for the departments of administration, financial planning, budgetary control, cost accounting, tax management payroll accounting, etc. and all personnel matters. He will be assisted in their duties by their respective assistants to assist in day to day activities.

Manpower planning and production

The central issue here may be one of scale. Production is staffed bye personnel, but the process may be labour-intensive or capital-intensive. Either way, planning will include:

1.   Analysis of labour supply and demand factors in relation to skills and training needs.
2.   Procedure for manpower recruitment, together with selection processes.

3.   The formation of industrial relations policies necessary for effective work place bargaining, disciplinary measures and dismissal procedures.
4.   Analysis of the effective use of human resources.
5.   Conditions necessary to maintain adequate levels of motivation. The scale of the problem is likely to be directly proportional to the method of production.

The availability of main persons is not going to be easy since TV industry is currently in its infancy in Africa.

Training needs

The selection and training of the required manpower for the proposed project has to be planned in advance.

The key personnel should be selected and trained

suitably. The training would be carried out in the following manner:

Basic training on the concept of TV industry before construction begins with visits to similar TV Stations in other countries. On site training during the construction phase of the project. On job training during the commissioning phase of the project. On job training during operation of the TV immediately after commissioning. The training of the key personnel such as Station Director should be carried out in all the phases. The training of other operating personnel should be suitably carried out during the construction, erection and operation phases in addition to training them by visits to similar plants operating in other countries.

Besides training the key operating staff described above, in TV training should also be given to other employees at skilled operating level to enable them to understand the process equipment in the project and prepare them to operate a maintain their respective sections safely, efficiently and skilfully. The above training should be carried out during construction, commissioning and operating phases of the project.

Training is necessary in order to enable personnel to acquire the skills and knowledge necessary to perform a task to an acceptable standard. The length of the training period and training methods will, of course, vary from job to job. Training is essentially a learning process, and in order that progress can be successfully monitored certain conditions are necessary.

1.   The training needs of both the individual and the organisation shall be identified and analysed.

2.   Targets and standards shall be set for the trainee, which are within his capabilities.

3.   The pace of the training programme should reflect the trainee's ability to maintain progress in properly absorbing the same.

4.   The trainee shall receive regular feedback of results. Any problem areas shall be highlighted, discussed and resolved.

5.   As the trainee progresses the amount of information provided shall be gradually reduced, thus inducing a feeling of independence and competence.

It is common place to find a wide variety of tasks in an organisation and each will require varying degrees of skill, effort and responsibility. This being so, it is inevitable that rates of pay will also vary and the differentials between the jobs will reflect their relative values. However, other factors such as local market conditions, bargaining strengths and traditions also influence a company's payment structure and a great deal of planning is required if rationalisation is to be achieved. One technique which has been successfully adopted by many companies to establish an equitable wage structure is job evaluation. In a job evaluation exercise a comparison is made of common criteria over a range of jobs, and the resulting analysis may be linked to a points allocation or job ranking system, and hence to a wage scale.

In conducting a job evaluation exercise it is important to cover a reasonable variety of tasks within the whole spectrum. For each, a job description is prepared setting out details of the duties and responsibilities undertaken by the employee together with a statement about his working conditions. Very often this task is undertaken by work study personnel since, they are responsible for determining methods of operation and evaluating the work content of the job. Each job is assessed factor by factor, resulting in a comprehensive comparative analysis.

The individual is the most important resources of any company and only people who are well trained, well motivated and adequately rewarded will provide a positive and synergistic contribution towards the company's objective and its organisational health.

In most cases, the factors, which may be weighted

according to relative value, are as follows:

1.   Skill-education, experience and training.
2.   Effort-both physical and mental.
3.   Responsibility-for equipment, materials, initiative etc.
4.   Working conditions-general conditions, risk of accident and injury.

Pay policies affect not only individual employees but the whole organisation, and the rewards and objectives vary at different levels within the enterprise.

Industrial relations

An effective industrial relations policy is important, since is the system through which employees take part in decision-making and in many instances it affects the while atmosphere of employer/employee relationships. An industrial relations policy is essentially a set of rules whose determine procedures for negotiation on such matters as:

1.   Wage and salary scales.
2.   Terms and conditions of work.
3.   Disputes and grievances.
4.   Recruitment and dismissal.
5.   Other issues of mutual interest, e.g. closed shop, redundancies and joint consultation.

In order to promote an atmosphere of co-operation, and to minimise conflict, the needs of management and work people must be recognised by both sides. Trade unions exit to protect the interests of their members and improve their working conditions. Management, while aware of the pressures and constraints imposed by the trade unions, have a duty to maximise the use of resources at their disposal, which may be expressed in relation to profitability, return on investment, level of service, sales volume, market share and cost-effectiveness. The strategies adopted in attempting to solve industrial relations problems will vary from company to company, and indeed from union to union, but there is no doubt that they will be influenced by both internal and external factors.

Internal factors

1.	The attitudes of employees to management, and management to employees.
2.	The leadership style of management.
3.	The bargaining strength of both parties.
4.	The number of negotiating bodies.
5.	The prosperity of the company.

External factors

1.	The extent to which parent boards influence company management, and district officials influence or control local shop stewards.

2.   Whether or not bargaining is conducted at plant, local or national level.

3.   Government policy towards industrial relations.

4.   The economic situations nationally, locally or within the company itself.

3.5   Describe the project implementation schedule.

Implementation of this project is a challenging task and calls for meticulous planning, scheduling and monitoring to realise the project goals within the budgeted cost and time frame. The goal can be achieved by adopting modern project management techniques.

To implement this project adequately, a team of engineers and project personnel having requisite education and experience are being appointed, to whom a detailed Work Breakdown Structure (WBS) in a logical order of activities, shall be supplied shortly, keeping in view cost estimation, scheduling, and to help monitor and control of the project. It is proposed to be formulated in conjunction with the objectives of each activity and goal settings. Project shall be programmed and controlled by network analysis techniques. Before the application of the network analysis techniques, the project personnel shall be acquainted with their capabilities in saving time, resources and costs. The training of project personnel and engineers at levels shall be provided for proper control of project progress and taking of timely corrective actions to re-align these efforts to meet ore stated objectives. It is proposed to gear programming and control system, i.e. project implementation system, which will ensure an integrated approach to project implementation. Project management activities shall be determined in advance and all activities carried out be project personnel as well as those to be contracted shall be identified.

Responsibility for project implementation shall be clearly defined. The forms of project organisation range from project oriented to functional organisation, while most of the cases are combinations of the two, with certain adaptations to prevailing conditions. It is impossible to over emphasise the importance of establishing a team of a task force for implementing the project with a designated leader to co-ordinate and guide its functions.

Project manager: Shall be responsible to the Board of Directors. The project manager shall be responsible for guiding and co-ordinating the efforts of all parties engaged in implementing the project, obtaining necessary government approvals on contracts. He is to control the project organisation with the promoters as well as with other agencies and organisations interested in the project. The manager shall have some staff to assist him, especially in checking expenditures to date and determining the present and future cost overrun or under run so that the project manager can take or propose to the Board pertinent corrective measures.

The network shall cover the pre-construction phase of the project indicating major administrative processes, since experience shows that some of them have frequently involved lengthy delays. In other words, it shall include the aggregate activities to be carried out be the principal parties participating in the implementation process.

The project budget shall then be prepared. Part or periodic payments to contractors which might be made at the end of certain intervals (e.g. weekly or monthly) throughout the time horizon of project implementation, shall be made by summing up activity costs per unit of time, which may be a week & month, and computing the cumulative cost at the end of each time interval. For the activities that are in process and are contacted or sub-contracted, the assumption of a linear time cost activity relationship shall be used for the sake of simplicity. In other words, expenditures are uniformly distributed throughout the duration of the activity.

PROJECT SCHEDULE

After an investment decision is taken, the main machinery and long delivery items must be ordered out at the earliest, forming the first major step in implementation of the project. It is foreseen that an engineering consultant will be appointed for carrying out the detailed engineering including basic engineering and procurement assistance to the client. It is also assumed that reputed and experienced contractors with adequate resources viz., men, materials, tools and tackles etc. will be engaged for execution of the construction and erection work. The purchase packages for auxiliaries shall be kept minimum so as to reduce the co-ordination efforts to the minimum. A great deal of co-ordination is required for constructing/erecting the new units. This task is feasible, provided the major activities of the project are co-ordinated and completed in the duration specified to achieve the respective milestones in time.

STRATEGY FOR TIMELY EXECUTION

It is important to deploy a team of experienced personnel for project execution and select the external agencies with due care for rendering the services and supply of equipment for the project. The project activities must be identified, planned and scheduled, and the progress monitored for timely project implementation. All the inputs to the project including financial resources must be identified and their inflow planned and arranged in time.

The project must be managed professionally with necessary co-ordination among the various agencies and requisite decisions taken promptly.

Establishment of an effective monitoring procedure for progress review and co-ordination.

In short, the following key factors would constitute the broad strategy for timely execution of all activities in a pre-determined manner as per schedule shown in the bar chart as to reach at a basis of regular production.

Early selection of an effective in-house technical team (TASK FORCE) by Government of Sao Tome, headed by a Project Manager for planning and executing the project.

i)   Proper choice of external agencies such as consultants for Project Engineering., Machinery suppliers, Construction Agencies etc. keeping in view their reputation/past performance and working experience in their fields.

ii)   Adequate use of computer-based PERT/ CPM techniques for project planning, scheduling and monitoring.

3.6 Description of the initial project Flowchart.

The project implementation phase embraces the period from the decision to start the project to the beginning of the commercial production. It includes a number of stages including negotiations and contracting, project design, construction and start-up.

3.7 Description of the requirement of Land.

For this project the Government of Sao Tome will participate with about 25.000 square meters of land. In selecting the land the following criteria should be followed.

1. The land should be near to the main city of Sao Tome.
2. The land should be properly connected by road.
3. It should also have proximity to many end user clients.
4. It should also meet various Government policy of
       achieving the social objectives.

2.08 Description of the requirement of Building.

The total requirement of building is as follows.

a. Adm. Building   400 Square Meters
b. Studios –3 (Three) 1500 Square Meters
c. Other Misc. Building   100 Square Meters

Total Square Meters: approximately 2000 Square Meters.
Cost for construction of about US$.750 per square meter (including air conditioning systems, primary electrical cabin etc.).

2.9   Description of some requirements for Plant & Machinery and their basis of selection.

The plant and machinery have been selected giving due consideration to the sophisticated nature of technology required. Detailed discussions were carried out with the foreign suppliers to ensure that the required capacities are practical with minimum capital and operating costs before the machinery were finally selected.

The machinery shall be selected from the most reputed foreign manufacturers of complete range of equipment. Keeping in view the following main factors:

a)   Past performance
b)   Existing machinery in Africa & abroad.
c)   Sales and service facilities in Africa

The main equipment is as follows.

STUDIO

1.   Post-production equipment – non-linear post-production, two ES7 stations.
2.   ENG – 2 DVCAM Camcorders
3.   Régie with 2 DXC-D30PK1 Digital Cameras - for cold light installation.
4.   Flexicart.
5.   Computer graphics suite.

Other technical data:
The whole installation is characterised by:
6. Informational web with UTP level, 5 cables.
7.   Phone system.
8.   Specific furniture.

Discrimination of Technical Data:

Studio
Cold lighting system.
2 Camera channels composed by:
1   DXC-D30PK Sony Digital Cameras 2 J18 Canon Zoom Lens
2 CA-537P Sony Camera Adapters

1   DXF-50 Sony Studio Viewfinder's
1   CCU-M3P Sony Control Camera Units 2 RM-
7P Sony remote CCU controls

2 50M Sony Multicore Camera Sets
2 Canon Focus and Zoom control sets
2 Intercommunication Headphones
2 Special tripods
1   charriot plane and curve
1   Teleprompter with 2 reading systems (monitors for individual speakers) Different microphones
with diverse directional characteristics
Audio components Video components

Video Régie

1 Recorder/Player BETACAM equipment
1 video digital mixer with 12 tracks
1 DVE equipment of digital effects for 3 channels
1 TBC Frame Synchroniser
1 Oscilloscope and Vectorscopy for control of cameras
1 Synchroniser Generators with Changeover
1 Intercommunication system with 4 places
1 Audio and Video Matrix with remote controls Video distributors

Audio Régie

1 audio mixer with 24 tracks

1 audio monitors
1 cd player
1 cassette deck
1   DAT
1   Mini-Disc
1   Level Detector for Audio Stereo Stereo Audio Distributors
Audio compressors Audio effects Microphones

Video Post-Production
2 Sony ES-7 on DVCAM/AVID hybrid non linear edition stations 1 Conventional BETACAM Edition Suite 2:1
Multi-track audio (8) on hard disc, Sound Scape type

3 ENG SETS:
Sony DSR-300 PK DVCAM Camcorder Digital Compact Report projectors
Various support materials (Batteries etc.)

3.10 Description of the Water Requirements for the project.

The requirements of water, separately for various matters are given in the following table.

Circulating   : Nil

Make-up   : Nil
Process   : Nil
Drinking   : 3,000 LPD

3.11 Description of the Steam requirement for the project.

A. Steam requirements and steam balance   : N.A.

B. Capacity and type of boiler with detailed specifications   : N.A.

C. Steam and energy diagram : N.A.

D.   Total energy generated / purchased (converted into M. K. : N.A. Cal) theoretical requirement of energy (in M. K. Cal)
at the various consumption stations and expected actual requirement at these stations.

E. If alternate processed are available, comparative energy : N.A.
consumption figures for the various processes. If the
project is energy intensive, possibility of choosing alternate process in order to make the project less energy
intensive.

F. Steps proposed to be taken by the company to improve   : N.A.

energy losses efficiency and reduce energy losses (such
as power factor improvement, power load management,
optimising, illumination waste heat utilisation, etc.)

G. Scope for usage of solar / other renewable sources of   : N.A.
energy.

H. Any other measures contemplated in the direction of : N.A.
energy conservation and management.

3.123 Some information on Compressed air, fuel, etc.

Compressed Air

(a) Requirement   : N.A.
(b) Sources   : N.A.
(c) Arrangements proposed   : N.A.
(d) Cost at site with detailed calculations : N.A.

3.13   Details of the nature of atmospheric, soil and water pollution likely to be created by the project and the measures proposed for control of pollution. Indicate whether necessary permissions for the disposal of effluent have been obtained.

There shall not be any atmospheric pollution likely to be created by the project as there is no machine which has combustion resulting into air pollution or any chemical process which may release any gases which may result into an air pollution.

However, the use of DG set shall cause a small amount of air pollution. Considering the size of DG set, the pollution is within the permissible limits.

SAO TOME TELEVISION NETWORK

THE INDUSTRY

CHAPTER - 4

THE INDUSTRY

4.01 Description of the Television Industry in general.

The television industry can be described in the following broad categories.

1. Introduction

2. Cable Networks

3. DTH

4. TV-media

5. Earnings drivers

6. Outlook

The detailed discussion in each of the categories is given below.

Introduction

The growing popularity of TV as a communication medium has resulted in the TV media sector undergoing a rapid transformation. From the black and white days of state controlled TV Station, to the highly colourful tunes of Channel V and MTV, the medium has certainly undergone a phenomenal change. Given its popularity, percentage ad spend has also increased proportionately on this medium.

Media pie (%)

	1995	1997
TV	62.5	68.8
Radio	20.9	15
Press	16.6	16.2

Source NRS

Entry of new channels post 1991

All over the world the telecasting has witnessed entry of new channels to cater to the various needs of world audiences. Channels have been launched in English as well as other regional languages. In many countries of the world till 1991, the state owned TV Station ruled the roost, as other players were not allowed to uplink and broadcast. However channels such as CNN, Star TV and BBC, which were offshore companies, could circumvent these regulations and telecast their programs into any country of the world. Cable operators then relayed the same and made it available to the common man through the cable television network.

Like many other countries, the State machinery controlled television. It was used as a propaganda tool for the party in power, with the opposition always at the receiving end. The customer had very little choice. The first steps towards more user choice began during the 1980s, which had to be telecast to a wider audience. TV Station used satellite channels for the telecast and the T.V. network was launched as an international channel.

The sports telecast by Channel 9 in 1985 and the Gulf War in the late eighties all played small but important cameos in educating the international viewer. With liberalisation in 1992 and crumbling tariff barriers televisions (read as colour TVs) became more easily available. The media revolution had started.

Major satellite channels avidly watched by viewers are Star TV, Sony TV, Home TV, BBC & CNN. There are other regional language channels which are major players in their respective territories.

Most of the channels that could not attain popularity rapidly suffered, since their advertisement earnings were not sustainable. The first round of the media wars is over. Management changes, i.e.

original promoters selling out to new management with deeper pockets, has become the order of the day. Alliances like the famous ESPN Star Sports arrangement also made headlines. Given the global trends of mergers and acquisitions, further consolidation is likely. Alliances and mergers make sense when the partners complement each other, like BBC and Discovery launched Animal Planet, CNBC and ABNI came together to launch a business channel called CNBC Asia.

Cable Networks

Antennas set up by either the end user or the cable operator receives the signals transmitted by the satellite. Local cable operators lay their own cables, set up control rooms, which can telecast 40 or more channels over a limited area. They charge the household a one- time connection charge of about US$.10 per point and a recurring monthly charge ranging between US$.1 to US$.5.

Initially, this was done in a very unorganised manner. The business required local knowledge and contacts, so every locality had its own cable operator. Collection was critical for the cable operator. For the end user, quality of telecast and a complete lack of standards became an issue. This lead to a shakeout and the formation of cable companies with money power which in turn tied-up with the local and small cable operators. Cable companies charge about US$1 per month to the local cable operators and support them with training and other infrastructure inputs. The business is immensely capital intensive and takes a long time to break even.

In many countries the operations of cable operators are regulated under the Cable TV Act which ensures that pornographic materials and other materials which are against culture and values or are detrimental to national interests do not get telecast. Recently this act has been amended to include foreign channels also.

Direct To Home

DTH is a new technology that circumvents the cable operators by directly delivering a bundle of channels to the end user. DTH involves transmission of encoded audio/ video signals (Ku band) via satellite. The end user needs an antenna to receive the signals and a decoder (set top box) to unscramble the encrypted signals. DTH services elsewhere in the world are Echostar and DirecTV (USA) and BskyB (Europe). Rupert Murdoch of Star TV fame owns BskyB.

The size of the antenna in DTH will be 1.5-2 ft in diameter, making it easy to install and transport. In conventional cable, since signals are in C band, an

8ft- diameter antenna is needed. The basic difference in the business model is the hardware costs in DTH. In a cable system, the user pays a one time connect fee and monthly rentals, while in DTH he has to invest in hardware.

The antenna will cost about US$200-300 and decoder will cost about US$200. The African viewer might be reluctant to incur such heavy installation costs. Quality of telecast in DTH is superior to Cable TV and viewer can receive up to 200 channels.

DTH will result in restructuring of the cable television industry. It will become imperative to have cash reserves to withstand the technology threat. Up gradation to fibre optic backbone will become necessary. A fibre optic network will cost about US$0.5mn per km as compared to US$0.1mn per km for coaxial cable. The stage is now ripe for consolidation.

TV Media features

In the Broadcasting business, it is only the industry leader who makes sizeable profits. The business is a game of asymmetrical payoffs. For instance, the top 5 channels account for 90% of ad spend.

Urbanisation and TV penetration is related. This may be due to the popularity of cable television that has resulted in increased colour TV sales. Rural penetration is low, although growing at a fast pace, because of dearth of specific program content to cater to that segment.

Liberalisation has resulted in the world viewer becoming more aware and conscious. This has resulted in the customer having more choice with the entry of a number of companies in different segments. Competition has resulted in companies increasing their marketing spend significantly.

Popularity of TV media is becoming higher. Increasing TV penetration leads to a reallocation of advertisement budgets with higher allocation for television at the cost of other medium.

TV channel operators use different business models to generate revenues. The critical component of any channel is the quality and type of programs they telecast. This determines their popularity, which in turn determines amount of advertisement revenues they can generate. They can do any one of the following:

Buy programming rights of program software from outside and collect advertisement revenue on their own. This model is followed by several TV companies, wherein they have a separate company in their fold, which develops all the content. The advantage is that re runs of serials/ programs become very profitable.

Selling time space to the producers for a fixed charge. Producers in turn are free to book advertisements at their own rates (there is an understanding on the time allocated for advertisement) and collect revenue. This is the basic model for many TV companies in which they sell prime time slots. The rights continue to be vested with the producer.

Earnings drives

The key factors that drive sector revenues are

Television penetration: Since the medium is television, increased television penetration will imply higher viewer ship. This will translate into higher advertisement spend allocation. This will also imply higher software production and demand for new programs.

Competition from other satellite channels would have an adverse impact on advertisement revenues, as advertisers have more choice in allocating ad budgets.

Government policies can have a big impact on the fortunes of the entire industry. When the DTH bill is passed in any country then, it will trigger a restructuring of the cable business.

Launching new channels targeted at specific segments, like regional channels within any country other areas having large pockets of ethnic population would lead to revenue growth. This will entail significant initial outlays.

Depreciation of the local currency would increase revenues as most of the program/ software companies export the programs overseas and payments are dollar denominated.

Advertisement revenue

As mentioned earlier, this is the primary source of income for TV channel operators. This revenue is directly co-related with the reach and viewer ship of a channel. Any channel's popularity depends on good quality programs, which is the software content. The business requires enormous initial investment in programs and revenues follow only with a time lag after the channel receives a minimum viewer acceptance.

Outlook

The sector has latent potential for growth on back of the exponential growth of cable TVs during the last 5 years. Television penetration in Africa is extremely low as compared to other developing countries like Malaysia, Pakistan, etc. in Asia. The number of channels has increased, implying higher demand for software programs.

Advertisement revenues, which are the barometer of channel popularity, will get dispersed over several competing channels. A shakeout is likely in both the channel and cable TV sectors. The biggest beneficiaries will be the content providers or the software houses. They will control the intellectual rights to the key element driving any channel's popularity.

Direct-to-Home, Digital Terrestrial Transmission and Conditional Access Cable Delivery have emerged as new delivery mechanisms. Breakthrough in technology would help open up avenues for these channels.

SAO TOME TELEVISION NETWORK

MARKETING

CHAPTER - 7

MARKETING

7.1   Description of the commercial viability of the project with regards to revenue generations.

Estimations shows that, after about 5 years of regular working, the CPLP Cultural Television Channel will be able to start exporting a great quantity of services and videos to Africa, Europe and South America. (See financial projections)

The economic feasibility of the project will occur with the commercialisation of several products, like:

1.   The sale and production of cultural programmes to various institutions like:
-    BBC
-    TV Culture Brazil
-    RTP
-    Discovery
Etc.

All over the world commercial networks for television cultural programmes have been created. Several television channels in different countries have regularly acquired programmes by producers spread out all over the world. Not only, several international organisations, like the World Bank and the United Nations, including FAO and Unesco, among many others, need programmes for public education, like programmes oriented to alert and to educate people concerning endemic and epidemic diseases. AIDS, malaria, tuberculosis are a very few examples.

These world institutions have made a great effort to develop television programmes with humanitarian objectives. But, in Africa – undoubtedly the continent most needed of such programmes – there is no television station, in present times, with capability to make face to such a need.

Therefore, the television teams that had been responsible for such programmes are, practically in its totality, placed in countries of the called First World, strange to the local population's concrete reality.

Programmes focusing new agricultural techniques or even oriented to agricultural, commercial and industrial education are essential elements of such a repressed demand.

The price for each institutional campaign, with an average of 10 films produced per campaign, is of about USD.15.000$ and the capacity of the CPLP Cultural Channel will be as follows:

First Group* - from 2 to 4 campaigns per month.
Second Group – from 1 campaign in two months to 1 campaign per month

This represents:
First Group   a potential annual income of about US$.540.000
Second Group a potential annual income of about US$.90.000

*First Group: Angola, Cape Verde, and Mozambique. Second Group: Guinea-Bissau, Sao Tome e Principe, East Timor (Lorosae). Reference Group: Brazil, Portugal.

2. The sale and production of no cultural programmes to other television channels, like:
-         series (novels)
-         talk-shows Etc.

Many countries of the region need to produce television programmes, but do not have capability to do it. Thus, they seem themselves obliged to search expensive productions in Europe. With only one programme sold-three daily hours-the channels of the First Group* would receive in incomes the equivalent of about two times of the whole investment. The Second Group countries will not have conditions at the beginning to product cultural programmes to other television channels.

*First Group: Angola, Cape Verde, and Mozambique. Second Grope: Guinea-Bissau, Sao Tome e Principe, East Timor (Lorosae). Reference Group: Brazil, Portugal.

The price for this type of television programme is:

Auditory programs   US\$.12.500 per programme
Talk-shows   US\$.4.000 per programme
Interviews   US\$.4.000 per programme

The capacity of production, of these programmes by the countries of the First Group will be:

In the FIRST PHASE:
Talk-shows   2 per month
Interviews   4 per month
In the FINAL PHASE:
Auditory programmes 6 per month
Talk-shows   20 per month
Interviews   20 per month

Thus, the potential income will be:

FIRST PHASE
Talk-shows   US$.96.000 per year
Interviews   US$.192.000 per year

FINAL PHASE
Auditory programmes   US$.900.000 per year
Talk-shows   US$.960.000 per year
Interviews   US$.960.000 per year

The price of the novels is much higher. For each complete novel, with about 120 chapters, the price is about US$.2.500.000 and the capacity of the First Group of the CPLP Cultural Channel (in the final phase) will be of one novel per year.

3. There is the possibility of commercial use of the transmission time beyond the six hours reserved to culture and education.

The conventional day in television is of 18 hours, of which only 6 hours would be no commercial. We would have, therefore, 12 hours of no cultural transmission, which should be freely commercialised.

Even the period of six hours- divided into two sections of three hours, the first one dedicated to culture and second section of three hours to education-has a great potential for sponsoring.

Each commercial hour of transmission can include up to 12 minutes of advertising. The price for each 30 seconds of advertising is:

Noble time   from 6PM to 11PM   a b o u t
US$.400 each 30"
Normal time   rest   about US$.200 each 30"

Concentrating the commercial programming in the noble time. With 4 hours of transmission in this period, we would have:

Educational television: from 7AM to 9AM
From 3PM to 4PM
Cultural television: from 4PM to 7PM
Commercial television: from 9AM to 3PM
From 7PM to 1AM

Thus, the commercial period would comprehend 6 hours in normal time, 4 hours in noble time and 2 hours in normal time again.

Begin 12 minutes per hour for advertising, the commercial period could have:
Noble time   48 minutes   24   films   US$.9.600
per day
Normal time   96 minutes   48   films   US$.9.600
per day
Total of the potential income............ US$.19.200

per day

> or US$.576.000 per month
> or US$.6.912.000 per year*

It is believed that in the first phase of the project (after one year), the CPLP Cultural Television Channels of the First Group will be able to start with an income from commercial advertisement of about US$.1.000.000 per year-value which is predicted to increase in the follow months.

Obs. * The amounts for Sao Tome are about US$.3.500.000 per year

4. The rent of the studios to television teams of other countries can be another source of incomes. Many producers who cover events in Africa need to move many times to their countries of origin during the video works, because there is not technical support in Africa. The same phenomenon happens with the cinematography and the journalism productions. It is not difficult to imagine, for example, the serious problems journalism teams have had, for example, with essential components like batteries, lighting, electronic components etc., which only can be easily find in Europe.

The period of renting of a television studio is of 12 hours. Each period has price of about US.4.000. After the first year, the studio should be rented for 12 hours per each period of 3 days. So, the rent per month, in this period (First Phase), will be able to generate incomes of about US$.120.000 per month in the First Group.

5. Support services to other television networks in all areas, including novels, mini series, docdramas series etc.

Many countries, principally in Africa, do not have technical conditions to develop this kind of programmes, but they have a strong internal repressed demand in this sense.

Each team, abroad, has a price of US$.1.000 per day (two people), after the costs of dislocation, hotel and meals. In the first months (first phase) it will be organised one team for works in other countries. In the final phase it is predicted to have up to 5 teams with such function. The capacity of each team is of about 20 days per month. Therefore, the support to other television networks-when took in its full potentiality-will be able to represent up to 20 days per month in the first phase. This represents a potential income of about US$.240.000 per year, and 100 days per month of works in the final phase, signifying about US$.1.200.000 per year of incomes, always referring to the First Group countries.

6. Colloquies, seminars, videoconferences and meetings of different natures.

The building of the CPLP Cultural Television Network will have a medium size auditorium, (First Group), with all conditions to receive seminars, colloquies and meetings of the most diverse nature. Such seminars, videoconferences etc. are important not only for the increase of the incomes, but also to attract specialists of the most different areas, being an important element for the development of all region as well as for the diffusion of the local, regional and continental cultural values.

The meetings, seminars and colloquies attract, in average, about 300 people per event. The price-excepting meals, hotels and transportation-per each participant is of about US$.50 per day. It will be a capacity for up to four events of this type per month, what could represent an income of about US$.60.000 per month or US$.720.000 per year (final phase-after 5 to 7 years).

The project should also turn possible.
-       classes of professional formation in diverse disciplines
-       support for a multimedia high technology centre (First Group)
-       support for a multimedia high cultural centre (Final Group)

SAO TOME TELEVISION NETWORK

PROFITABILITY & CASH FLOW

CHAPTER - 8

PROFITABILITY AND CASH FLOW

8.1   Estimation of cost of production and working results for the first five years of operation.

The estimated cost of production of working results
for the first five years of operation are given in the
chapter "Financial Projection".

8.2   Cash flow statement for the company as a
whole, for five operating years of the project based
on the estimates of working results.

A detailed cash flow statement for the company as
a whole for five operating years is given in the chapter of "Financial Projection".

8.3   Projected balance sheet for five operating years for the company as a whole.

The balance sheet for five operation years for the company as a whole is given in the chapter of "Financial Projection".

SAO TOME TELEVISION NETWORK

ASSUMPTIONS

CHAPTER - 9

ASSUMPTIONS

1.   The Sao Tome Television Network shall generate the income from the following sources.

a)   Sale of cultural programmes.

b)   Sale of commercial programmes.

c)   Sale of advertisement time during transmission of TV programmes

d)   Hiring of studios.

e)   Supply of technical services

f)   Hiring of conference hall

The detailed assumptions for each of the above mentioned activities are as follows:

a)   Sale of cultural programs:

I.   It is assumed that four cultural programs shall be produced per month in the first year, which can be sold to other countries. This
figure will increase to 5 programs per month in the second year, six programs per month in the third year and so on.

II.   It is assumed that the selling price shall be US$.75 per program.

b)   Sale of commercial programs:

I.   The commercial programs consist of talk shows, interviews and auditory programs.

II.   It is assumed that two talk shows shall be produced per month in the first year, three talk shows shall be produced in the third year, four talk shows per month shall be produced in the third year and
so on.

III.   It is assumed that the selling price of one talk show program shall
be US$.2000.

IV.   It is assumed that four interviews shall be produced per month in the first year, five interviews shall be produced in the third year, six interviews per month shall be produced in the third year and so on.

V.   It is assumed that the selling price of one interviews program shall be US$.2000.

VI.   It is assumed that no auditory programs shall be produced in the first and the second year. Only when the people have two years of experience then in the third year one auditory program shall be produced per month. In the fourth year two programs shall be produced per month and so on.

VII.   It is assumed that the selling price of one auditory program shall be UD$.6.250.

c) Sale of advertisement time during the transmission of TV programs:

I.   There shall be a total transmission of 8 hours per day in the first year. It will go on increasing by two hours in the second and subsequent years.

II.   Out of which advertisement shall be available for programs with transmission period of four hours in the first year. The programs in which advertisements could be available shall increase by two hours per year in the second and subsequent years.

III.   In one hour T.V. transmission, 12 minutes of advertisement shall be allowed.

IV.   Out of total T.V. time 33 % time shall be considered as prime time and remaining 67% shall be considered as non prime time, in the first, second and third years. From the fourth year onwards the prime-time advertisements shall remain constant.

V.   The advertisement rates shall be US$ 400 for one minute of prime time advertisement and US$ 175 for one minute of non prime time advertisement.

d)   Hiring of studios:

I.   It is assumed that the studios shall be taken on hire for a shift of 12 hours, two times in a month in the first year. It will increase to
three times in a month in the second year.

II.   The rent per day shall be US$.2.000.

e)   Supply of technical services:-

I.   It is assumed that one team, consisting of two technically qualified people shall be available in the first year. In the second year two

such teams shall be available. In the third year three such teams shall be available and so on.

II.   It is assumed that in the first year the team

shall be hired for 10 days in a month. In the second year the teams shall be hired for 12

days in a month. In the third year the teams shall be hired for 14 days in a month and so on.

III.   The rate of hire shall be US$.500 per team per day.

f)   Hiring of conference hall :-

I.   It is assumed that the conference hall shall be taken on hire for two times in a month in the first year, three times in a month in the second year and four times in a month in the third year and so on.

II.   The hire charges shall be US$.250 per conference.

2.   The unit is planning transmission of three hours of educational programs, three hours of cultural programs and two hours of commercial programs per day in the first year. Thus, there shall be a total transmission of eight hours of programs per day in the first year. The total transmission hours shall increase by two hours per day in the second and subsequent years. Thus, in the second year there shall be total transmission of 10 hours per day and in the third year there shall be a total transmission of 12 hours per day. The cost of software production is assumed at US$.1.250 per hour.

3.   The cost of miscellaneous consumable

items is assumed at US$.18.000 in the first year. This will increase by 10% in the second and subsequent years.

4.   The cost of stores and spares consumed is assumed at US$.21.000 in the first year. It will increase by 10% in the second and subsequent years.

5.   The cost of repairs and maintenance is assumed at 0.1% of the total cost of plant and machinery in the first year. This cost will increase up to 0,3% in the fifth year.

6.   It is assumed that the unit shall use 1.000 units of power every day or 365.000 units of power in the first year. The consumption of power shall increase by 10% in the second and subsequent years. The cost of power is assumed at 2.5 cents for one unit.

7.   It is assumed that the unit shall consume 1.000 litters of diesel per month. The cost is assumed at 16.5 cents for one litter. It will increase by 10% in second and subsequent years.

8.   The unit shall have a technical staff of 35 people and administrative staff of 15 people.

9.   Depreciation is calculated on reducing bal-

ance method. The rate of depreciation is taken at 10% for building, 25% for plant and machinery, 10% for furniture and fixtures and 20% on vehicles.

10.   Advertisement expenditure is assumed at 0.2% of the sales.

11.   It is assumed that the stock of raw material and stores shall be 30 days of consumption, receivables shall be 45 days sales and creditors shall be 15 days purchases.

12.   The term loan shall be for a total period of 25 years. It shall have a moratorium of five years. It shall be repayable in 20 equal yearly instalments.

TELEVISÃO SONHO
Rede Planetária de Televisão Cultural para a Paz Mundial

PROJECT REPORT

OF

TIMOR LOROSAE CULTURAL TELEVISION NETWORK

AT

TIMOR LOROSAE

ASIA

PROMOTERS

REPUBLIC OF TIMOR LOROSAE

ASIA

TELEVISÃO SONHO
Rede Planetária de Televisão Cultural para a Paz Mundial

TIMOR LOROSAE TELEVISION NETWORK

INTRODUCTION

CHAPTER - 1

INTRODUCTION

1.1   Description of the organisation through which this project is being taken up.

The CPLP is an international organisation, created in 17 of July of 1996, with Headquarters in Lisbon and consisting of the following State Members: Angola, Brazil, Cape-Verde, Guinea-Bissau, Mozambique, Portugal, Sao Tome e Principe and East Timor (Timor Lorosae).

The CPLP has as its objective the politic-diplomatic relationships between its Members, mainly referring to the United Nations and the World Bank, the relationship EU/MERCOSUL and the accomplishment of the Europe-Africa Summit.

The CPLP is also dedicated to the co-operation, particularly in the economic, social, cultural, legal and technical-scientific matters; as well as to the projects of promotion and broadcasting of the Portuguese Language, nominated to the improvement of the International Institute of the Portuguese Language and the creation of a Bibliographical Fund.

All its decisions are taken by consensus.

2.2   Description of the main reasons for setting up an African and Asian Cultural Television Network.

Africa is a continent submerged into dramatic problems. All these humanitarian problems have in the education and culture their most direct and objective root.

The African countries speaking Portuguese are among the African countries with the lowest level of development.

However Timor Lorosae is not located in Africa, but in Asia, it is member of the CPLP and its situation, given the recent political events, puts the country in the front line for the Cultural Television Network Project.

The whole situation is so dramatic that the implementation of the CPLP Cultural Television Network is more than urgent.

It is hoped that with this Television Network the

people shall have better understanding of various diseases and other problems of life. e.g. AIDS is prevalent disease in Africa. But as the people are illiterate the information about the disease cannot be spread by any other method. In such cases the Television Network becomes a very effective audio-visual means of communication.

The project also intends to be a base of a massive production of health campaigns, specially to rein-force the combat against the AIDS, the infect-con-tagious diseases- endemic or epidemic-through a process of global education of the local and region-al populations.

The CPLP Cultural Television Network is a project for peace and its objective is not to compete with local television channels already installed, but to interchange with them, promoting a future global web dedicated to knowledge and development.

1.03 Description of the main objects of the project.

The major objective of the project is to assist the CPLP countries to develop the first African televi-sion network channel oriented to cultural and edu-cational themes with the follow orientation:

-        diffusion of the Portuguese language
-        education in all areas

- reinforcement of the local and regional cultures
- formation of technical personal on communication and tele-education

- formation and information for a better participation of the CPLP nations in the process of global development
- to support the implantation of the Institute of Portuguese Language

The development of such a project will create an irradiation pole from the countries with the project implementation to the whole African continent, constituting an African network of research and culture, discovering and valorising different cultural aspects and, consequently, reinforcing identity factors.

With the stronger cultural identity, it is expected to have lower levels of violence in social behaviour as well as a relevant increase of the educational performance in all its sectors.

Systematic educational objectives are complementary to the general orientation of this project, which is cultural in its fundament.
A very large positive impact is also to be expected with respect to several sectors of the economy, greatly improving the overall quality of services.

1.4   Description of the mission status of the project.

The mission of the project is:

-       to rescue lost values and cultural information
-       to reinforce historical elements
-       to rescue old artworks
-       to give a new value to new artworks
-       to communicate practical and objective information on health
-       to communicate practical and objective information on agriculture
-       to communicate practical and objective information on economy

1.05 Description of the philosophy of the project.

The CPLP Cultural Television Network will have in its philosophy, as a support to its mission, the follow fundaments:

-       to work with local people
-       be focused on quality
-       intercommunication

Three words can define the CPLP Cultural Television Network philosophy:

People, quality and intercommunication

Following these three key words of its philosophy, a single sentence also defines its mission: "A Continent of Culture".

Two main sectors characterise its global structure:
1. Cultural Sector
1.1.   To develop programmes on:
-       Music
-       Architecture (also vernacular)
-       Archaeology
-       Literature and poetry
-       History
-       Tourism
-       Dance
-       Plastic arts
-       Astronomy
-       Science in general
-       gastronomy
-       health
-       anthropology
-       local cultures
-       others

1.2   The programmes can be of all types, the most important is to be attentive to communication. All programmes must reach the highest possible number of spectators.

2. Educational sector.

2.1   Systematic education in short practical courses.
2.2   Educational series with practical and objective information on:
-       health care
-       family
-       agriculture
-       managing
-       services
-       civil construction

As the CPLP Cultural Television Network must not be supported by the respective governments in the future, it will have a daily time of free commercial programmes up to 12 hours. This is a good solution to pay its costs and support its development.

1.06 Description the magnitude of the project.

The project proposes a Television transmission time of eight hours per day in the first year. It will increase by two hours per day in the second and subsequent years.

The project also envisages training of local people in all the related areas of Television. It is proposed to develop a local pool of talent which can be utilised not only in this particular country but also in other countries of Africa.

The total cost of the project of the Timor Lorosae

Cultural Television Network is US$.10.600.000. Out of this it is proposed to obtain a soft loan of US$.8.800.000. The remaining amount of US$.1.800.000 shall be contributed by the Government of Timor Lorosae in terrain for the installation of the respective building, infrastructures and logistic support.

For the purpose of the Timor Lorosae Cultural Television Network a company shall be formed with the participation of the Government. The Contribution of the Government in the form of share capital shall be US$.1.800.000.

The term loan shall be on soft terms. The total period of the loan shall be 23 years. Out of which five years shall be a period of moratorium. The entire loan shall be repaid in the remaining 17 years, in 17 equal yearly instalments.

The Government of Timor Lorosae will give the sovereign guarantee for the referred loan.

TIMOR LOROSAE TELEVISION NETWORK

THE COUNTRY

CHAPTER - 2

THE COUNTRY

2.1 - Description the level of development of Timor Lorosae.

East Timor, occupying the eastern part of the island of Timor, is situated between Australia and Indonesia.

Following a referendum in August 1999, East Timor was placed under the supervision of the United Nations. On 9 December 1999, a Trust Fund for East Timor (TFET) was established with the assistance of funding agencies. The TFET grant projects are formulated and supervised by ADB and the World Bank.

The economy was severely affected by the violence and destruction that followed the referendum, with some estimates indicating a 40 percent decline in real gross domestic product (GDP) in 1999. Public and private property was damaged, the agricultural cycle and trade were disrupted, and a substantial part of infrastructure was destroyed. Low cash activities; low demand; and lack of capital, infrastructure, and a transport system constrain the economy. Unemployment is high and humanitarian assistance continues to be an important source of support for a great part of the population.

As conditions in the country gradually returned to normal, the economy rebounded. Given the low base, real GDP growth was estimated at 15 percent in 2000, largely as a result of reconstruction work and the gradual restoration of commerce, business, and basic services. Because of favourable weather, the output of maize, rice, and coffee (the major export commodity) also improved in 2000. Inflation slowed, along with reduced regional disparities in staple food prices.

The recovery process is expected to gather momentum and be led by agriculture, construction, and services; prices are expected to stabilise in the coming years. The combined sources budget of East Timor for FY2001 (ending 30 June) provides for a total revenue of $25.1 million and total expenditure, including net lending, of $183.0 million. The overall deficit of $157.9 million (about 55 percent of GDP) will be financed largely from external grants.

In 2000, the National Consultative Council adopted the US dollar as the legal tender, although the Indonesian rupiah continued to be used. The Central Payments

Office began providing basic depository and payment services. Several regulations related to banking and the payments system were enacted. The Central Fiscal Authority has also started revenue and budgetary functions. The Department of Economic Affairs and the National Planning and Development Agency were created.

The intense level of violence that followed the 30 August referendum resulted in a massive population displacement, both within and across the borders of East Timor. Food distribution and marketing systems, together with commerce and essential services, were paralysed and the agricultural cycle was disrupted. In addition, although infrastructure and property damage was extensive, agricultural damage was less severe as crops had already been harvested and only limited burning of the relatively minor second season crop in fields, occurred.

The main damage was to livestock, food and seed stocks, which were either looted or burnt, and to disruption in the agricultural cycle, which either delayed or will delay key operations such as land preparation and planting. The main constraint to planting the main crop of maize, in November/December, is population displacement, which will mean that in some areas a relatively large proportion of farmers will not have returned to complete sowing. In addition, although the availability of maize seeds was earlier envisaged as a serious impediment, international efforts to distribute seeds, through the

UN and NGOs, have been highly effective, covering a large proportion of requirements.

As the remainder will come from farmer exchanges and use of stored seeds that were not destroyed, the loss in production in 2000 will be lower than thought earlier. As rice planting/transplanting does not need to be completed till mid January, there is greater scope for recovery. Population displacement and seed shortages are not considered to be major problems for the current rice season, though there will be some reduction in area planted due to the marked reduction in animal and mechanised traction, which are essential for land preparation. Although the rainy season started somewhat late, rainfall during November/early December was generally favourable, benefiting crops.

Although the level of disruption in agriculture was less than anticipated at the height of the crisis, there will inevitably be a notable decline in cereal production, given the constraints outlined above.

East Timor is known to have pockets of malnutrition but it is not well documented. The El Niño-related drought, which seriously affected food production in 1998, and the financial crisis in Asia, which resulted in tremendous price hikes, greatly exacerbated problems of food availability over the last year.

East Timor is the eastern half (approximately) of the

island of Timor. It is a small mountainous territory, around 14 500 km2, in the Indian Ocean some 500 km to the north of Australia. Even before recent events and population displacement, East Timor was among the poorer regions in Asia. Approximately 50 percent of the population are considered below the poverty line, life expectancy is around 56 years, whilst only two out five people are literate. The level of infrastructure and the provision of essential services are also poor, with only 30 percent of households having access to potable water and 22 percent having electricity. Less than half (49 percent) of the number of villages were accessible by paved road before the crisis.

Although the environment and the natural resource base of the island are extremely important in development they are not as conducive as in many other small island states. The economy is primarily agricultural, contributing the largest share to GDP, employing almost three quarters of the workforce, providing over 70 percent of the population with their main sources of livelihood and offering the greatest potential for exports and trade. The sector, in turn, is largely dominated by subsistence production of primary staples, (maize, rice, cassava and sweet potatoes), employing family labour and little purchased inputs.

Land use, however, is highly constrained by rugged topography, poor soils, erosion and unpredictable rainfall. In general, soils do not support heavy vegetation, which limits the scope for crop diversification. Because of these constraints, producers are largely risk averse, employing a strategy aimed at minimising potential losses rather than maximising yields. Periodically the island is affected by El Niño related weather anomalies, which further reduce food production. It is estimated that around 600 000 hectares are suitable for agriculture of which approximately half are currently being used.

In addition around 200 000 hectares are suitable for rearing livestock, which, hitherto, were the territory's second largest export after coffee. However, after the recent events and given security, ownership concerns and lack of capital investment, the sector remains well below potential.

Although fisheries offer considerable scope, a very small proportion of potential is currently exploited as the industry remains largely under-developed. Less than half the work force in fisheries is full time, and the equipment used inefficient. The main fishing region is off the coast around the capital Dili and Atauro.

The potential for developing forestry is limited due to severe deforestation mainly connected with slash and burn agricultural practices. In general there is little scope for further development, as only small tracts are suitable for expansion. Hitherto, sandalwood was an important and valuable export but over-exploitation led to the cessation of exports in 1930s.

The climate is dominated by intense north-east monsoon rainfall from November to March, followed by a pronounced dry season. In northern parts of the island, this constitutes the main wet season, whilst in the south, the rainy season is more prolonged spanning December to June, increasing the potential for secondary crop production of maize and possibly rice. The northern coast has the lowest average rainfall, ranging from 600mm to 1 000 mm per annum. In contrast, total precipitation is highest in central mountainous areas with a range from 1 800 to 3 000mm. Rainfall in southern coastal areas ranges from around 1 250 to 3 000mm. See figure 1, for an indication of rainfall distribution.

On 5 May 1999, agreement was reached between the Governments of Indonesia and Portugal for phased transition in the territory. The three phases agreed encompassed: (1) the period between 5 May 1999 and 30 August 1999 for preparation and execution of a referendum on independence (2) the post referendum phase, during which Indonesia would retain responsibility for security and Government services and (3) the transition to independence overseen by a UN-led transitional civil and peace keeping authority. Contrary to the agreement aimed at peaceful transition, however, widespread civil unrest and violence followed the August 30 referendum, in which the majority (78 percent) of Timorese rejected autonomy in favour of independence.

Although precise figures are still unknown, many people were killed in the violence, whilst hundreds of thousands were internally displaced or exiled as refugees to West Timor. In addition, enormous damage was inflicted on infrastructure, essential services and property. Available estimates, for example, indicate that Dili, the capital, and smaller urban centres were extensively damaged with up to 50 percent of social infrastructure and buildings destroyed. Moreover, individual household assets (primarily livestock and basic goods), were also destroyed or looted. Overall, therefore, in the intense wave of post referendum violence, essential fabrics of the economy and commerce were almost entirely demolished.

Although the situation has improved significantly following Security Council Resolution 1264 and the deployment of an international peace keeping force (INTERFET), the reconstruction of East Timor will require considerable time and resources, both capital and human. In the interim, humanitarian assistance will be vital to safeguard food security and the provision of basic needs.

In addition to agricultural and property damage,

East Timor lost essential human resources as educated leaders and service staff were specifically targeted in the violence, while the majority of civil servants left the province as many were Indonesian. The gap in administration, service and skilled personnel will invariably have an adverse affect on economic functions in the short to medium term.

TIMOR LOROSAE TELEVISION NETWORK

THE PROJECT

CHAPTER - 3

THE PROJECT

3.1   Description of the initial steps to be taken for the implementation of the project.

The first objective of the CPLP Cultural Television

Network is to install its structures in the following countries:

- Sao Tome e Principe
- Cape Verde
- Mozambique
- Guinea-Bissau
- Angola
- Timor

Thus, these six countries will structure the basic network of the whole project.

Angola, Mozambique and Cape Verde will form a FIRST GROUP; Sao Tome e Principe, Guinea-Bissau and East Timor will form a SECOND GROUP; and, finally, Brazil and Portugal will form a called REFERENCE GROUP.

However, the project will start at the same time in all countries. A more precise description of the meaning of each group is made below.

The basic physical components of the project are:

1.   The creation of companies, in each of the States of the CPLP, oriented to the objectives above referred.

2.   The State of each country – who will be the responsible and guarantee of the payment of the financing – will also have a participation in the com-

panies.

3.	All strategies will be oriented to lowest costs in long term.

4.	The whole project will be developed by phases.

5.	The timetable of transmission will be defined case by case.

6.	The project will include:

6.1.	a complete plan for an equipment network.

6.2.	a complete plan for a software network.

6.3.	all procurements

6.4.	equipment transportation and installation

6.5.	all configurations

6.6.	creation of the infographic system

6.7.	technical formation of the personal

6.8.	architecture and engineer projects for the building

6.9.	construction of the building organised in phases

1.2	Description of the procedure by which the necessary programmes will be produced for transmission on the channel.

The television programs are known as software in this field. The unit is planning transmission of three hours of educational programs, three hours of cultural programs and two hours of commercial programs per day in the first year. Thus, there shall be a total transmission of eight hours of programs per day in the first year. The total transmission hours shall increase by two hours per day in the second and subsequent years. Thus, in the second year there shall be total transmission of 10 hours per day and in the third year there shall be a total transmission of 12 hours per day. The cost of software production is assumed at US$.2.500 per hour. The amount of US$.2.500 per hour is taken based on cost of production of such programmes in Portugal.

To contract a similar system from other television channel, only for one country, for a transmission period of 6 hours daily, the cost in, market (eg. Portugal*) is of about US$.500.000 per month. This cost represents a total of about US$.6.000.000 per year. In 16 years, excluding the first year of implementation, the costs for contract the six daily hours of transmission from other television channel, in present market price, is of about US$.100.000.000. The six countries would have an estimated cost of about US$.600.000.000 in a 16 years period of time.

These costs are based on the prices presently used in Portugal, as for example the programme Hora Viva – Seguranca em Directo, transmitted every day, excluding weekends, from 7 to 10 AM.

Another example is the channel Canal Noticias Lisboa CNL. Only in its first year of installation, it was expended about US$.10.000.000, and the equipment as well as the building were not property of the channel, but rented.

In fact, however the project will be implemented by phases, after 5 to 10 years of regular work, it is predicted to have reached a similar value in incomes as the examples above.

*value took in the Portuguese market, below the average of the European prices in about 30%.

The first ten months of works will be oriented to:
-        design of the equipment and software structures
-        start the first productions
-        elaboration of the projects of architecture and engineering
-        procurement and acquisition of all equipment
-        construction and supervision of the works
-        installation and configuration of the equipment

So, the first year must be used to the implantation of the project. However, as to accelerate the chronogram, some operations should be started before this timing, temporarily installed in a different building.

The first phase will be oriented to:

- video productions (external and internal)
- edition
- elaboration of cost programmes

3.3 Description of the requirements of Power, Fuel and Water. POWER

It is assumed that the unit shall use one thousand units of power of everyday. Thus, the unit shall use 365.000 units of power in a year. The costs of power is assumed at two cents for one unit. The consumption of power shall increase by 10% every year.

FUEL

The unit shall use D.G. set for generation of electricity, when there is an interruption in the supply. The hours of disruption in the normal supply can not be predicted accurately. Hence, the hours for which the D.G. sets shall be used also can not be predicted accurately. However, it is assumed that the unit shall utilise one thousand litters of diesel per month. The cost is assumed in about 13 cents for one litter. Hence, the total cost for one thousand litters shall be US$.130. The consumption of diesel shall increase by 10% every year.

WATER

The unit does not need water for fixed commercial

operation. However, it shall have a strength of 50 people as employees. Water shall be required for drinking and sanitation purposes. It is assumed that 3000 litters of water shall be required everyday. The cost of obtaining this water is assumed at US$.4 everyday. The consumption of water shall increase by 10% every year.

3.4   Description of the manpower requirement of the project and description of the steps taken by the company to train the manpower.

The total staff strength shall be 17 people in the first year.

The system of modern TV Station is designed to achieve sustained operating efficiency and transmission. This, of course, entails a certain degree of sophistication in the production and transmission. However, considering the socio-economic situation in Timor Lorosae, a reasonable balance has to be struck to obtain optimum performance and at the same time create gainful employment. While working out the manpower requirement for this project to be kept on direct rolls of the company totalling 50 staff and operators, the above consideration have been kept in mind.

Manpower requirement

The direct manpower required for the proposed Unit is about US$.720.000 per year. The manpower requirement as indicated in this chapter has been planned keeping in view the following guide-lines:

Effective co-ordination among the various departments. Judicious distribution of responsibilities.

Capacity utilisation of the TV Station with optimum manpower.

Details of manpower requirement for the TV Station is given in the financial section.

In line with the prevailing practice, the Security guards, office peons and unskilled labour etc., are normally employed on contractual/casual basis, and their cost has to be included in the Factory overheads. However in this report provision has been made to employ the above staff and their salaried have been included.

The various departments proposed shall be under the direct responsibility of the Station Director. He shall be assisted by Deputy Directors, who will look after the complete TV Station and its various day to day activities and Marketing. They shall be responsible for achieving the envisaged targets and sales forecasts. They shall be assisted by a team of Managers from production, marketing and accounts.

When evaluating an investment project from an employment point of view, its impact on both unskilled and skilled labour has been taken into account. Not only direct employment, but also indirect employment has been considered. Direct employment refers to the new employment opportunities created within the project; indirect employment concerns job opportunities created in other projects linked with the project which is being formulated.

The implementation of large and sophisticated projects generally contributes to the development of local skills and capabilities in a country. Furthermore, they help to change traditional values, attitudes and the behaviour of the society, to build up an enterprising spirit among the people, to develop a desire for changing and improving the existing conditions of life, to introduce better work discipline and thus to change the very pattern and basis of economic development. The TV industry is already well established in Africa. Location of TV industry activities in Africa, has certain favourable factors and advantages, in setting up this type of industry, as availability of acceptable level of education among the supervisory staff adequate technical and managerial skills developed over a long period, and availability of cheap labour are assured.

The organisation structure will vary from TV Station to TV Station in the industry and as such the pattern proposed herein can be considered only as suggestive and provisional.

A well-knit organisation structure headed by a Station Director and Manager, with the supporting staff will be developed progressively during project implementation. Soon after the plant becomes operative, a good number of project staff will be absorbed in the organisation. Certain additional staff also get added to ensure smooth and efficient management of the operating unit.

The Manager Production will have a degree in communication and will be accountable and answerable to the Station Director for all TV operations including planning, production, material management, TV utilities, quality control, production cost and budgetary control, TV safety, discipline and layout relations. His main function should be to ensure achievement for quality and quantity targets of production at reasonable cost and should constantly strive to improve TV performance. In the discharge of his multifarious duties and responsibilities he will be assisted by supervisors and adequate staff for day to day activities.

His duties are to ensure that targets are maintained through effective utilisation of 3 M's viz: Machinery, Men and Materials. These targets should be translated in terms of targets for individuals under him such as supervisors, skilled workers etc. He has to ensure that the equipment at his disposal gets prompt attention on breakdown is adhered to strictly. He has to further ensure that there is a strict quality control exercised over raw materials also. He will be assisted by Deputy Manager etc. who and will report to the General Manager.

GENERAL MANAGER (Comm.) He will report to the Station Director. His main duties will be prepare sales forecasts and budgets, study and monitor the export market for the company's products and advice on ways and means of increasing sales. He will also ensure that consumer complains are solved in an appropriate manner. He will also ensure that the right materials are available for production at the right time.

He will be responsible for all aspects connected with the export procedures and keep the management updated on all matters relating to the Govt. policies on polymers and Exports and above all the world market.

COMPANY SECRETARY & FINANCIAL CONTROLLER. He will report to the Station Director and will be responsible for the departments of administration, financial planning, budgetary control, cost accounting, tax management payroll accounting, etc. and all personnel matters. He will be assisted in their duties by their respective assistants to assist in day to day activities.

Manpower planning and production

The central issue here may be one of scale. Production is staffed bye personnel, but the process may be labour-intensive or capital-intensive. Either way, planning will include:

1.   Analysis of labour supply and demand factors in relation to skills and training needs.
2.   Procedure for manpower recruitment, together with selection processes.
3.   The formation of industrial relations policies necessary for effective work place bargaining, disciplinary measures and dismissal procedures.
4.   Analysis of the effective use of human resources.
5.   Conditions necessary to maintain adequate levels of motivation. The scale of the problem is likely to be directly proportional to the method of production.

The availability of main persons is not going to be easy since TV industry is currently in its infancy in Africa.

Training needs

The selection and training of the required manpower for the proposed project has to be planned in advance.

The key personnel should be selected and trained suitably. The training would be carried out in the following manner:

Basic training on the concept of TV industry before construction begins with visits to similar TV Stations in other countries. On site training during the construction phase of the project. On job training during the commissioning phase of the project. On job training during operation of the TV immediately after commissioning. The training of the key personnel such as Station Director should be carried out in all the phases. The training of other operating personnel should be suitably carried out during the construction, erection and operation phases in addition to training them by visits to similar plants operating in other countries.

Besides training the key operating staff described above, in TV training should also be given to other employees at skilled operating level to enable them to understand the process equipment in the project and prepare them to operate a maintain their respective sections safely, efficiently and skilfully. The above training should be carried out during construction, commissioning and operating phases of the project.

Training is necessary in order to enable personnel to acquire the skills and knowledge necessary to perform a task to an acceptable standard. The length of the training period and training methods will, of course, vary from job to job. Training is essentially a learning process, and in order that progress can be successfully monitored certain conditions are necessary.

1. The training needs of both the individual and the organisation shall be identified and analysed.

2. Targets and standards shall be set for the trainee, which are within his capabilities.

3. The pace of the training programme should reflect the trainee's ability to maintain progress in properly absorbing the same.

4. The trainee shall receive regular feedback of results. Any problem areas shall be highlighted, discussed and resolved.

5. As the trainee progresses the amount of information provided shall be gradually reduced, thus inducing a feeling of independence and competence.

It is common place to find a wide variety of tasks in an organisation and each will require varying degrees of skill, effort and responsibility. This being so, it is inevitable that rates of pay will also vary and the differentials between the jobs will reflect their relative values. However, other factors such as local market conditions, bargaining strengths and traditions also influence a company's payment structure and a great deal of planning is required if rationalisation is to be achieved. One technique which has been successfully adopted by many companies to establish an equitable wage structure is job evaluation. In a job evaluation exercise a comparison is made of common criteria over a range of jobs, and the resulting analysis may be linked to a points allocation or job ranking system, and hence to a wage scale.

In conducting a job evaluation exercise it is important to cover a reasonable variety of tasks within the whole spectrum. For each, a job description is prepared setting out details of the duties and responsibilities undertaken by the employee together with a statement about his working conditions. Very often this task is undertaken by work study personnel since, they are responsible for determining methods of operation and evaluating the work content of the job. Each job is assessed factor by factor, resulting in a comprehensive comparative analysis.

The individual is the most important resources of any company and only people who are well trained, well motivated and adequately rewarded will provide a positive and synergistic contribution towards the company's objective and its organisational health.

In most cases, the factors, which may be weighted according to relative value, are as follows:

1. Skill-education, experience and training.
2. Effort-both physical and mental.
3. Responsibility-for equipment, materials, initiative etc.
4. Working conditions-general conditions, risk of accident and injury.

Pay policies affect not only individual employees but the whole organisation, and the rewards and objectives vary at different levels within the enterprise.

Industrial relations

An effective industrial relations policy is important, since is the system through which employees take part in decision-making and in many instances it affects the while atmosphere of employer/employee relationships. An industrial relations policy is essentially a set of rules whose determine procedures for negotiation on such matters as:

1. Wage and salary scales.
2. Terms and conditions of work.
3. Disputes and grievances.
4. Recruitment and dismissal.

5.   Other issues of mutual interest, e.g. closed shop, redundancies and joint consultation.

In order to promote an atmosphere of co-operation, and to minimise conflict, the needs of management and work people must be recognised by both sides. Trade unions exit to protect the interests of their members and improve their working conditions. Management, while aware of the pressures and constraints imposed by the trade unions, have a duty to maximise the use of resources at their disposal, which may be expressed in relation to profitability, return on investment, level of service, sales volume, market share and cost-effectiveness. The strategies adopted in attempting to solve industrial relations problems will vary from company to company, and indeed from union to union, but there is no doubt that they will be influenced by both internal and external factors.

Internal factors

1.   The attitudes of employees to management, and management to employees.
2.   The leadership style of management.
3.   The bargaining strength of both parties.
4.   The number of negotiating bodies.
5.   The prosperity of the company.

External factors

1.   The extent to which parent boards influence company management, and district officials influence or control local shop stewards.

2.   Whether or not bargaining is conducted at plant, local or national level.

3.   Government policy towards industrial relations.

4.   The economic situations nationally, locally or within the company itself.

3.5   Describe the project implementation schedule.

Implementation of this project is a challenging task and calls for meticulous planning, scheduling and monitoring to realise the project goals within the budgeted cost and time frame. The goal can be achieved by adopting modern project management techniques.

To implement this project adequately, a team of engineers and project personnel having requisite education and experience are being appointed, to whom a detailed Work Breakdown Structure (WBS) in a logical order of activities, shall be supplied shortly, keeping in view cost estimation, scheduling, and to help monitor and control of the project. It is proposed to be formulated in conjunction with the objectives of each activity and goal settings. Project shall be programmed and controlled by network analysis techniques. Before the application of the network analysis techniques, the project personnel shall be acquainted with their capabilities in saving time, resources and costs. The training of project personnel and engineers at levels shall be provided for proper control of project progress and taking of timely corrective actions to re-align these efforts to meet ore stated objectives. It is proposed to gear programming and control system, i.e. project implementation system, which will ensure an integrated approach to project implementation. Project management activities shall be determined in advance and all activities carried out be project personnel as well as those to be contracted shall be identified.

Responsibility for project implementation shall be clearly defined. The forms of project organisation range from project oriented to functional organisation, while most of the cases are combinations of the two, with certain adaptations to prevailing conditions. It is impossible to over emphasise the importance of establishing a team of a task force for implementing the project with a designated leader to co-ordinate and guide its functions.

Project manager: Shall be responsible to the Board of Directors. The project manager shall be responsible for guiding and co-ordinating the efforts of all parties engaged in implementing the project, obtaining necessary government approvals on contracts. He is to control the project organisation with the promoters as well as with other agencies and organisations interested in the project. The manager shall have some staff to assist him, especially in checking expenditures to date and determining the present and future cost overrun or under run so that the project manager can take or propose to the Board pertinent corrective measures.

The network shall cover the pre-construction phase of the project indicating major administrative processes, since experience shows that some of them have frequently involved lengthy delays. In other words, it shall include the aggregate activities to be carried out be the principal parties participating in the implementation process.

The project budget shall then be prepared. Part or periodic payments to contractors which might be made at the end of certain intervals (e.g. weekly or monthly) throughout the time horizon of project implementation, shall be made by summing up activity costs per unit of time, which may be a week & month, and computing the cumulative cost at the end of each time interval. For the activities that are in process and are contacted or sub-contracted, the assumption of a linear time cost activity relationship shall be used for the sake of simplicity. In other words, expenditures are uniformly distributed throughout the duration of the activity.

PROJECT SCHEDULE

After an investment decision is taken, the main machinery and long delivery items must be ordered out at the earliest, forming the first major step in implementation of the project. It is foreseen that an engineering consultant will be appointed for carrying out the detailed engineering including basic engineering and procurement assistance to the client. It is also assumed that reputed and experienced contractors with adequate resources viz., men, materials, tools and tackles etc. will be engaged for execution of the construction and erection work. The purchase packages for auxiliaries shall be kept minimum so as to reduce the co-ordination efforts to the minimum. A great deal of co-ordination is required for constructing/erecting the new units. This task is feasible, provided the major activities of the project are co-ordinated and completed in the duration specified to achieve the respective milestones in time.

STRATEGY FOR TIMELY EXECUTION

It is important to deploy a team of experienced personnel for project execution and select the external agencies with due care for rendering the services and supply of equipment for the project. The project activities must be identified, planned and scheduled, and the progress monitored for timely project implementation. All the inputs to the project including financial resources must be identified and their inflow planned and arranged in time.

The project must be managed professionally with necessary co-ordination among the various agencies and requisite decisions taken promptly.

Establishment of an effective monitoring procedure for progress review and co-ordination.

In short, the following key factors would constitute the broad strategy for timely execution of all activities in a pre-determined manner as per schedule shown in the bar chart as to reach at a basis of regular production.

Early selection of an effective in-house technical team (TASK FORCE) by Government of Timor Lorosae, headed by a Project Manager for planning and executing the project.

i)   Proper choice of external agencies such as consultants for Project Engineering., Machinery suppliers, Construction Agencies etc. keeping in view their reputation/past performance and working experience in their fields.

ii)   Adequate use of computer-based PERT/ CPM techniques for project planning, scheduling and monitoring.

3.6   Description of the initial project Flowchart.

The project implementation phase embraces the period from the decision to start the project to the beginning of the commercial production. It includes a number of stages including negotiations and contracting, project design, construction and start-up.

3.7   Description of the requirement of Land.

For this project the Government of Timor Lorosae will participate with about 25.000 square meters of land. In selecting the land the following criteria should be followed.

1.   The land should be near to the main city of Dili.
2.   The land should be properly connected by road.
3.   It should also have proximity to many end user clients.
4.   It should also meet various Government policy of
        achieving the social objectives.

2.08 Description of the requirement of Building.

The total requirement of building is as follows.
a. Adm. Building   400 Square Meters
b. Studios –2 (Two)   1000 Square Meters
c. Other Misc. Building   200 Square Meters

Total Square Meters: approximately 1600 Square Meters.
Cost for construction of about US$.750 per square meter (including air conditioning systems, primary electrical cabin etc.).

2.9   Description of some requirements for Plant & Machinery and their basis of selection.

The plant and machinery have been selected giving due consideration to the sophisticated nature of technology required. Detailed discussions were carried out with the foreign suppliers to ensure that the required capacities are practical with minimum capital and operating costs before the machinery were finally selected.

The machinery shall be selected from the most reputed foreign manufacturers of complete range of equipment. Keeping in view the following main factors:

a) Past performance
b) Existing machinery in Africa & abroad.
c) Sales and service facilities in Africa

The main equipment is as follows.

STUDIO

1. Post-production equipment – non-linear post-production, two ES7 stations.
2. ENG – 2 DVCAM Camcorders
3. Régie with 2 DXC-D30PK1 Digital Cameras - for cold light installation.
4. Flexicart.
5. Computer graphics suite.

Other technical data:
The whole installation is characterised by:
1. Informational web with UTP level, 5 cables.
2. Phone system.
3. Specific furniture.

Discrimination of Technical Data:

Studio

Cold lighting system.
2 Camera channels composed by:
1   DXC-D30PK Sony Digital Cameras 2 J18 Canon Zoom Lens
2   CA-537P Sony Camera Adapters
1   DXF-50 Sony Studio Viewfinder's
1   CCU-M3P Sony Control Camera Units 2 RM-7P Sony remote CCU controls
2 50M Sony Multicore Camera Sets
2 Canon Focus and Zoom control sets
2 Intercommunication Headphones
2 Special tripods
1   charriot plane and curve
1   Teleprompter with 2 reading systems (monitors for individual speakers) Different microphones with diverse directional characteristics

Audio components Video components

Video Régie
1 Recorder/Player BETACAM equipment
1 video digital mixer with 12 tracks
1 DVE equipment of digital effects for 3 channels
1 TBC Frame Synchroniser
1 Oscilloscope and Vectorscopy for control of cameras
1 Synchroniser Generators with Changeover
1 Intercommunication system with 4 places
1 Audio and Video Matrix with remote controls Video distributors

Audio Régie

1 audio mixer with 24 tracks
1 audio monitors
1 cd player
1 cassette deck
1   DAT
1   Mini-Disc
1   Level Detector for Audio Stereo Stereo Audio Distributors

Audio compressors Audio effects Microphones

Video Post-Production
2 Sony ES-7 on DVCAM/AVID hybrid non linear edition stations 1 Conventional BETACAM Edition Suite 2:1
Multi-track audio (8) on hard disc, Sound Scape type

3 ENG SETS:
Sony DSR-300 PK DVCAM Camcorder Digital Compact Report projectors
Various support materials (Batteries etc.)

3.10   Description of the Water Requirements for the project.

The requirements of water, separately for various matters are given in the following table.

Circulating   : Nil
Make-up   : Nil
Process   : Nil
Drinking   : 3,000 LPD

3.11 Description of the Steam requirement for the project.

A. Steam requirements and steam balance   : N.A.

B. Capacity and type of boiler with detailed specifications   : N.A.

C. Steam and energy diagram : N.A.

D.   Total energy generated / purchased (converted into M. K. : N.A. Cal) theoretical requirement of energy (in M. K. Cal)

at the various consumption stations and expected actual requirement at these stations.

E. If alternate processed are available, comparative energy : N.A.

consumption figures for the various processes. If the
project is energy intensive, possibility of choosing alternate process in order to make the project less energy
intensive.

F. Steps proposed to be taken by the company to improve   : N.A.
energy losses efficiency and reduce energy losses (such
as power factor improvement, power load management,
optimising, illumination waste heat utilisation, etc.)

G. Scope for usage of solar / other renewable sources of   : N.A.
energy.

H. Any other measures contemplated in the direction of : N.A.
energy conservation and management.

3.123 Some information on Compressed air, fuel, etc.

Compressed Air

(a) Requirement   : N.A.
(b) Sources   : N.A.

(c) Arrangements proposed   : N.A.
(d) Cost at site with detailed calculations : N.A.

3.13   Details of the nature of atmospheric, soil and water pollution likely to be created by the project and the measures proposed for control of pollution. Indicate whether necessary permissions for the disposal of effluent have been obtained.

There shall not be any atmospheric pollution likely to be created by the project as there is no machine which has combustion resulting into air pollution or any chemical process which may release any gases which may result into an air pollution.

However, the use of DG set shall cause a small amount of air pollution. Considering the size of DG set, the pollution is within the permissible limits.

TIMOR LOROSAE TELEVISION NETWORK

THE INDUSTRY

CHAPTER - 4

THE INDUSTRY

4.01 Description of the Television Industry in general.

The television industry can be described in the following broad categories.

1.   Introduction

2.   Cable Networks

3.   DTH

4.   TV-media

5.   Earnings drivers

6.   Outlook

The detailed discussion in each of the categories is given below.

Introduction

The growing popularity of TV as a communication medium has resulted in the TV media sector undergoing a rapid transformation. From the black and white days of state controlled TV Station, to the highly colourful tunes of Channel V and MTV, the medium has certainly undergone a phenomenal change. Given its popularity, percentage ad spend has also increased proportionately on this medium.

Media pie (%)

	1995	1997
TV	62.5	68.8
Radio	20.9	15
Press	16.6	16.2

Source NRS

Entry of new channels post 1991

All over the world the telecasting has witnessed entry of new channels to cater to the various needs of
world audiences. Channels have been launched in
English as well as other regional languages. In many
countries of the world till 1991, the state owned TV
Station ruled the roost, as other players were not
allowed to uplink and broadcast. However channels
such as CNN, Star TV and BBC, which were offshore
companies, could circumvent these regulations
and telecast their programs into any country of the
world. Cable operators then relayed the same and
made it available to the common man through the
cable television network.

Like many other countries, the State machinery
controlled television. It was used as a propaganda
tool for the party in power, with the opposition always at the receiving end. The customer had very
little choice. The first steps towards more user
choice began during the 1980s, which had to be telecast to a wider audience. TV Station used satellite
channels for the telecast and the T.V. network was
launched as an international channel.

The sports telecast by Channel 9 in 1985 and the

Gulf War in the late eighties all played small but important cameos in educating the international viewer. With liberalisation in 1992 and crumbling tariff barriers televisions (read as colour TVs) became more easily available. The media revolution had started.

Major satellite channels avidly watched by viewers are Star TV, Sony TV, Home TV, BBC & CNN. There are other regional language channels which are major players in their respective territories.

Most of the channels that could not attain popularity rapidly suffered, since their advertisement earnings were not sustainable. The first round of the media wars is over. Management changes, i.e. original promoters selling out to new management with deeper pockets, has become the order of the day. Alliances like the famous ESPN Star Sports arrangement also made headlines. Given the global trends of mergers and acquisitions, further consolidation is likely. Alliances and mergers make sense when the partners complement each other, like BBC and Discovery launched Animal Planet, CNBC and ABNI came together to launch a business channel called CNBC Asia.

Cable Networks

Antennas set up by either the end user or the cable operator receives the signals transmitted by the satellite. Local cable operators lay their own cables, set up control rooms, which can telecast 40 or more channels over a limited area. They charge the household a one- time connection charge of about US$.10 per point and a recurring monthly charge ranging between US$.1 to US$.5.

Initially, this was done in a very unorganised manner. The business required local knowledge and contacts, so every locality had its own cable operator. Collection was critical for the cable operator. For the end user, quality of telecast and a complete lack of standards became an issue. This lead to a shakeout and the formation of cable companies with money power which in turn tied-up with the local and small cable operators. Cable companies charge about US$1 per month to the local cable operators and support them with training and other infrastructure inputs. The business is immensely capital intensive and takes a long time to break even.

In many countries the operations of cable operators are regulated under the Cable TV Act which ensures that pornographic materials and other materials which are against culture and values or are detrimental to national interests do not get telecast. Recently this act has been amended to include foreign channels also.

Direct To Home

DTH is a new technology that circumvents the cable operators by directly delivering a bundle of channels to the end user. DTH involves transmission of encoded audio/ video signals (Ku band) via satellite. The end user needs an antenna to receive the signals and a decoder (set top box) to unscramble the encrypted signals. DTH services elsewhere in the world are Echostar and DirecTV (USA) and BskyB (Europe). Rupert Murdoch of Star TV fame owns BskyB.

The size of the antenna in DTH will be 1.5-2 ft in diameter, making it easy to install and transport. In conventional cable, since signals are in C band, an 8ft- diameter antenna is needed. The basic difference in the business model is the hardware costs in DTH. In a cable system, the user pays a one time connect fee and monthly rentals, while in DTH he has to invest in hardware.

The antenna will cost about US$200-300 and decoder will cost about US$200. The African viewer might be reluctant to incur such heavy installation costs. Quality of telecast in DTH is superior to Cable TV and viewer can receive up to 200 channels.

DTH will result in restructuring of the cable television industry. It will become imperative to have cash reserves to withstand the technology threat. Up gradation to fibre optic backbone will become necessary. A fibre optic network will cost about US$0.5mn per km as compared to US$0.1mn per km for coaxial cable. The stage is now ripe for consolidation.

TV Media features

In the Broadcasting business, it is only the industry leader who makes sizeable profits. The business is a game of asymmetrical payoffs. For instance, the top 5 channels account for 90% of ad spend.

Urbanisation and TV penetration is related. This may be due to the popularity of cable television that has resulted in increased colour TV sales. Rural penetration is low, although growing at a fast pace, because of dearth of specific program content to cater to that segment.

Liberalisation has resulted in the world viewer becoming more aware and conscious. This has resulted in the customer having more choice with the entry of a number of companies in different segments. Competition has resulted in companies increasing their marketing spend significantly.

Popularity of TV media is becoming higher. Increasing TV penetration leads to a reallocation of advertisement budgets with higher allocation for television at the cost of other medium.

TV channel operators use different business models to generate revenues. The critical component of any channel is the quality and type of programs they telecast. This determines their popularity, which in turn determines amount of advertisement revenues they can generate. They can do any one of the following:

Buy programming rights of program software from outside and collect advertisement revenue on their own. This model is followed by several TV companies, wherein they have a separate company in their fold, which develops all the content. The advantage is that re runs of serials/ programs become very profitable.

Selling time space to the producers for a fixed charge. Producers in turn are free to book advertisements at their own rates (there is an understanding on the time allocated for advertisement) and collect revenue. This is the basic model for many TV companies in which they sell prime time slots. The rights continue to be vested with the producer.

Earnings drives

The key factors that drive sector revenues are

Television penetration: Since the medium is television, increased television penetration will imply higher viewer ship. This will translate into higher advertisement spend allocation. This will also imply higher software production and demand for new programs.

Competition from other satellite channels would have an adverse impact on advertisement revenues, as advertisers have more choice in allocating ad budgets.

Government policies can have a big impact on the fortunes of the entire industry. When the DTH bill is passed in any country then, it will trigger a restructuring of the cable business.

Launching new channels targeted at specific segments, like regional channels within any country other areas having large pockets of ethnic population would lead to revenue growth. This will entail significant initial outlays.

Depreciation of the local currency would increase revenues as most of the program/ software companies export the programs overseas and payments are dollar denominated.

Advertisement revenue

As mentioned earlier, this is the primary source of

income for TV channel operators. This revenue is directly co-related with the reach and viewer ship of a channel. Any channel's popularity depends on good quality programs, which is the software content. The business requires enormous initial investment in programs and revenues follow only with a time lag after the channel receives a minimum viewer acceptance.

Outlook

The sector has latent potential for growth on back of the exponential growth of cable TVs during the last 5 years. Television penetration in Africa is extremely low as compared to other developing countries like Malaysia, Pakistan, etc. in Asia. The number of channels has increased, implying higher demand for software programs.

Advertisement revenues, which are the barometer of channel popularity, will get dispersed over several competing channels. A shakeout is likely in both the channel and cable TV sectors. The biggest beneficiaries will be the content providers or the software houses. They will control the intellectual rights to the key element driving any channel's popularity.

Direct-to-Home, Digital Terrestrial Transmission and Conditional Access Cable Delivery have emerged as new delivery mechanisms. Breakthrough in technology would help open up avenues for these channels.

TIMOR LOROSAE TELEVISION NETWORK

MARKETING

CHAPTER - 7

MARKETING

7.1   Description of the commercial viability of the project with regards to revenue generations.

Estimations shows that, after about 5 years of regular working, the CPLP Cultural Television Channel will be able to start exporting a great quantity of services and videos to Africa, Europe and South

America. (See financial projections)

The economic feasibility of the project will occur with the commercialisation of several products, like:

1.   The sale and production of cultural programmes to various institutions like:
-       BBC
-       TV Culture Brazil
-       RTP
-       Discovery
Etc.

All over the world commercial networks for television cultural programmes have been created. Several television channels in different countries have regularly acquired programmes by producers spread out all over the world. Not only, several international organisations, like the World Bank and the United Nations, including FAO and Unesco, among many others, need programmes for public education, like programmes oriented to alert and to educate people concerning endemic and epidemic diseases. AIDS, malaria, tuberculosis are a very few examples.

These world institutions have made a great effort to develop television programmes with humanitarian objectives. But, in Africa – undoubtedly the continent most needed of such programmes – there is no television station, in present times, with capability to make face to such a need.

Therefore, the television teams that had been responsible for such programmes are, practically in its totality, placed in countries of the called First World, strange to the local population's concrete reality.

Programmes focusing new agricultural techniques or even oriented to agricultural, commercial and industrial education are essential elements of such a repressed demand.

The price for each institutional campaign, with an average of 10 films produced per campaign, is of about USD.15.000$ and the capacity of the CPLP Cultural Channel will be as follows:

First Group* - from 2 to 4 campaigns per month.
Second Group – from 1 campaign in two months to 1 campaign per month

This represents:
First Group   a potential annual income of about US$.540.000
Second Group a potential annual income of about US$.90.000

*First Group: Angola, Cape Verde, and Mozambique. Second Group: Guinea-Bissau, Sao Tome e Principe, East Timor (Lorosae). Reference Group:

Brazil, Portugal.

Obs. For the case of Guinea-Bissau and East Timor, it is expected a decrease of such incomes for US$.216.000 and US$.36.000.

2. The sale and production of no cultural programmes to other television channels, like:
- series (novels)
- talk-shows Etc.

Many countries of the region need to produce television programmes, but do not have capability to do it. Thus, they seem themselves obliged to search expensive productions in Europe. With only one programme sold-three daily hours-the channels of the First Group* would receive in incomes the equivalent of about two times of the whole investment. The Second Group countries will not have conditions at the beginning to product cultural programmes to other television channels.

*First Group: Angola, Cape Verde, and Mozambique. Second Grope: Guinea-Bissau, Sao Tome e Principe, East Timor (Lorosae). Reference Group: Brazil, Portugal.

The price for this type of television programme is:

Auditory programs   US$.12.500 per programme
Talk-shows   US$.4.000 per programme

Interviews   US$.4.000 per programme

The capacity of production, of these programmes by the countries of the First Group will be:

In the FIRST PHASE:
Talk-shows   2 per month
Interviews   4 per month
In the FINAL PHASE:
Auditory programmes 6 per month
Talk-shows   20 per month
Interviews   20 per month

Thus, the potential income will be:

FIRST PHASE
Talk-shows   US$.96.000 per year
Interviews   US$.192.000 per year

FINAL PHASE
Auditory programmes US$.900.000 per year
Talk-shows   US$.960.000 per year
Interviews   US$.960.000 per year

Obs. All values must be decreased for Guinea-Bissau and East Timor in the first years, as it is referred in the detailed feasibility studies.

The price of the novels is much higher. For each complete novel, with about 120 chapters, the price is about US$.2.500.000 and the capacity of the First

Group of the CPLP Cultural Channel (in the final phase) will be of one novel per year.

3. There is the possibility of commercial use of the transmission time beyond the six hours reserved to culture and education.

The conventional day in television is of 18 hours, of which only 6 hours would be no commercial. We would have, therefore, 12 hours of no cultural transmission, which should be freely commercialised.

Even the period of six hours- divided into two sections of three hours, the first one dedicated to culture and second section of three hours to education-has a great potential for sponsoring.

Each commercial hour of transmission can include up to 12 minutes of advertising. The price for each 30 seconds of advertising is:

Noble time   from 6PM to 11PM   a b o u t US$.400 each 30"
Normal time   rest   about US$.200 each 30"

Concentrating the commercial programming in the noble time. With 4 hours of transmission in this period, we would have:

Educational television: from 7AM to 9AM

From 3PM to 4PM
Cultural television: from 4PM to 7PM
Commercial television: from 9AM to 3PM
From 7PM to 1AM

Thus, the commercial period would comprehend 6 hours in normal time, 4 hours in noble time and 2 hours in normal time again.

Begin 12 minutes per hour for advertising, the commercial period could have:

Noble time   48 minutes   24 films US$.9.600
per day
Normal time   96 minutes   48 films US$.9.600
per day
Total of the potential income........... US$.19.200
per day
                     or US$.576.000 per month
                     or US$.6.912.000 per year

It is believed that in the first phase of the project (after one year), the CPLP Cultural Television Channels of the First Group will be able to start with an income from commercial advertisement of about US$.1.000.000 per year-value which is predicted to increase in the follow months.

4. The rent of the studios to television teams of other countries can be another source of incomes.

Many producers who cover events in Africa need to move many times to their countries of origin during the video works, because there is not technical support in Africa. The same phenomenon happens with the cinematography and the journalism productions. It is not difficult to imagine, for example, the serious problems journalism teams have had, for example, with essential components like batteries, lighting, electronic components etc., which only can be easily find in Europe.

The period of renting of a television studio is of 12 hours. Each period has price of about US.4.000. After the first year, the studio should be rented for 12 hours per each period of 3 days. So, the rent per month, in this period (First Phase), will be able to generate incomes of about US$.120.000 per month in the First Group.

5. Support services to other television networks in all areas, including novels, mini series, docdramas series etc.

Many countries, principally in Africa, do not have technical conditions to develop this kind of programmes, but they have a strong internal repressed demand in this sense.

Each team, abroad, has a price of US$.1.000 per day (two people), after the costs of dislocation, hotel and meals. In the first months (first phase) it will be organised one team for works in other countries. In the final phase it is predicted to have up to 5 teams with such function. The capacity of each team is of about 20 days per month. Therefore, the support to other television networks-when took in its full potentiality-will be able to represent up to 20 days per month in the first phase. This represents a potential income of about US$.240.000 per year, and 100 days per month of works in the final phase, signifying about US$.1.200.000 per year of incomes, always referring to the First Group countries.

6. Colloquies, seminars, videoconferences and meetings of different natures.

The building of the CPLP Cultural Television Network will have a medium size auditorium, (First Group), with all conditions to receive seminars, colloquies and meetings of the most diverse nature. Such seminars, videoconferences etc. are important not only for the increase of the incomes, but also to attract specialists of the most different areas, being an important element for the development of all region as well as for the diffusion of the local, regional and continental cultural values.

The meetings, seminars and colloquies attract, in average, about 300 people per event. The price-excepting meals, hotels and transportation-per each participant is of about US$.50 per day. It will be a capacity for up to four events of this type per month, what could represent an income of about US$.60.000 per month or US$.720.000 per year (final phase-after 5 to 7 years).

The project should also turn possible.

-	classes of professional formation in diverse disciplines
-	support for a multimedia high technology centre (First Group)
-	support for a multimedia high cultural centre (Final Group)

TIMOR LOROSAE TELEVISION NETWORK

PROFITABILITY & CASH FLOW

CHAPTER - 8

PROFITABILITY AND CASH FLOW

8.1   Estimation of cost of production and working results for the first five years of operation.

The estimated cost of production of working results
for the first five years of operation are given in the
chapter "Financial Projection".

8.2   Cash flow statement for the company as a
whole, for five operating years of the project based
on the estimates of working results.

A detailed cash flow statement for the company as
a whole for five operating years is given in the chapter of "Financial Projection".

8.3   Projected balance sheet for five operating
years for the company as a whole.

The balance sheet for five operation years for the company as a whole is given in the chapter of "Financial Projection".

TIMOR LOROSAE TELEVISION NETWORK

ASSUMPTIONS

CHAPTER - 9

ASSUMPTIONS

1.   The Timor Lorosae Television Network shall generate the income from the following sources.

a)   Sale of cultural programmes.
b)   Sale of commercial programmes.
c)   Sale of advertisement time during transmission of TV programmes

d)   Hiring of studios.
e)   Supply of technical services
f)   Hiring of conference hall

The detailed assumptions for each of the above mentioned activities are as follows:

a)   Sale of cultural programs:

I.   It is assumed that four cultural programs shall be produced per month in the first year, which can be sold to other countries. This

figure will increase to 5 programs per month in the second year, six programs per month in the third year and so on.

II.   It is assumed that the selling price shall be US$.60 per program.

b)   Sale of commercial programs:

I.   The commercial programs consist of talk shows, interviews and auditory programs.

II.   It is assumed that two talk shows shall be produced per month in the first year, three talk shows shall be produced in the third year, four talk shows per month shall be produced in the third year and so on.

III.   It is assumed that the selling price of one talk show program shall be US$.1600.

IV.   It is assumed that four interviews shall be produced per month in the first year, five interviews shall be produced in the third year, six interviews per month shall be produced in the third year and so on.

V.   It is assumed that the selling price of one interviews program shall be US$.1600.

VI.   It is assumed that no auditory programs

shall be produced in the first and the second year. Only when the people have two years of experience then in the third year one auditory program shall be produced per month. In the fourth year two programs shall be produced per month and so on.

VII.   It is assumed that the selling price of one auditory program shall be UD$.5.000.

c) Sale of advertisement time during the transmission of T.V. programs:

I.   There shall be a total transmission of 8 hours per day in the first year. It will go on increasing by two hours in the second and subsequent years.

II.   Out of which advertisement shall be available for programs with transmission period of four hours in the first year. The programs in

which advertisements could be available shall increase by two hours per year in the second and subsequent years.

III.   In one hour T.V. transmission, 12 minutes of advertisement shall be allowed.

IV.   Out of total TV time 33% time shall be considered as prime time and remaining 67% shall be considered as non prime time, in the first, second and third years. From the fourth year onwards the prime-time advertisements shall remain constant.

V.   The advertisement rates shall be US$ 320 for one minute of prime time advertisement and

US$ 140 for one minute of non prime time adver-
tisement.

d)   Hiring of studios:

I.   It is assumed that the studios shall be taken
on hire for a shift of 12 hours, two times in a month
in the first year. It will increase to three times in a
month in the second year.
II.   The rent per day shall be US$.1.600.

e)   Supply of technical services:-

I.   It is assumed that one team, consisting of
two technically qualified people shall be available
in the first year. In the second year two such teams
shall be available. In the third year three such teams
shall be available and so on.
II.   It is assumed that in the first year the team
shall be hired for 10 days in a month. In the second year the teams shall be hired for 12 days in a
month. In the third year the teams shall be hired
for 14 days in a month and so on.
III.   The rate of hire shall be US$.400 per team
per day.

f)   Hiring of conference hall :-

I.   It is assumed that the conference hall shall
be taken on hire for two times in a month in the first
year, three times in a month in the second year and

four times in a month in the third year and so on.

II.   The hire charges shall be US$.200 per conference.

2.   The unit is planning transmission of three hours of educational programs, three hours of cultural programs and two hours of commercial programs per day in the first year. Thus, there shall be a total transmission of eight hours of programs per day in the first year. The total transmission hours shall increase by two hours per day in the second and subsequent years. Thus, in the second year there shall be total transmission of 10 hours per day and in the third year there shall be a total transmission of 12 hours per day. The cost of software production is assumed at US$.1.000 per hour.

3.   The cost of miscellaneous consumable items is assumed at US$.14.400 in the first year. This will increase by 10% in the second and subsequent years.

4.   The cost of stores and spares consumed is assumed at US$.16.800 in the first year. It will increase by 10% in the second and subsequent years.

5.   The cost of repairs and maintenance is assumed at 0.1% of the total cost of plant and machinery in the first year. This cost will increase up to 0,3% in the fifth year.

6.   It is assumed that the unit shall use 1.000 units of power every day or 365.000 units of power in the first year. The consumption of power shall increase by 10% in the second and subsequent years. The cost of power is assumed at 2 cents for one unit.

7.   It is assumed that the unit shall consume 1.000 litters of diesel per month. The cost is assumed at 13 cents for one litter. It will increase by 10% in second and subsequent years.

8.   The unit shall have a technical staff of 35 people and administrative staff of 15 people.

9.   Depreciation is calculated on reducing balance method. The rate of depreciation is taken at 10% for building, 25% for plant and machinery, 10% for furniture and fixtures and 20% on vehicles.

10.   Advertisement expenditure is assumed at 0.2% of the sales.

11.   It is assumed that the stock of raw material and stores shall be 30 days of consumption, receivables shall be 45 days sales and creditors shall be 15 days purchases.

12.   The term loan shall be for a total period of 25 years. It shall have a moratorium of five years. It shall be repayable in 20 equal yearly instalments.

breve bio

Emanuel Dimas de Melo Pimenta (1957) tem sido considerado por muitos como um dos mais interessantes arquitetos, músicos, fotógrafos e artistas intermédia no início do terceiro milênio – de acordo com testemunhos escritos por John Cage, Ornette Coleman, Lucrezia De Domizio, Merce Cunningham, John Archibald Wheeler, René Berger, Dove Bradshaw, Daniel Charles, Phill Niblock ou William Anastasi entre outros.

Os seus trabalhos estão incluídos em algumas das mais importantes coleções de arte e em instituições mundialmente reconhecidas como o Whitney Museum de Nova Iorque; o museu de arte contemporânea ARS AEVI, de Sarajevo; a Bienal de Veneza; o CyberArt Museum de Seattle; o Kunsthaus de Zurique, a coleção de arte contemporânea Durini; a Biblioteca Nacional de Paris; e o MART Museu de Arte Moderna de Rovereto e Trento entre outros.

Ele desenvolve arquitetura, urbanismo e música usando tecnologias de realidade virtual, ciberespaço e neurociências.

Os seus trabalhos estão incluídos na *Enciclopédia Universalis* (*Enciclopédia Britânica*) desde

1991; no *Sloninky Baker's Music Dictionary* (Berkeley); na *Cronologia da Música Clássica Ocidental* de Charles Hall; no *All Music Guide – The Expert's Guide to the Best Cds*, na *Wikipedia* e no *Babilon* entre outros. Artigos sobre os seus trabalhos têm aparecido regularmente em diferentes jornais e revistas como *The Wire, Ear, The New York Times, Le Monde, le Parisien, Liberation, O Estado de São Paulo, O Expresso, O Público, O Globo, Il Sole 24 Ore* e *la Reppublica* entre outros.

No início dos anos 1980, Emanuel Pimenta cunhou o conceito de *arquitetura virtual*, mais tarde amplamente usado como disciplina em universidades de todo o mundo. Entre os seus desafiadores projetos há a famosa ilha flutuante para Lisboa, Portugal; uma ilha flutuante experimental para o Lago Maggiore, na Suíça; o edifício Symmetrion em Budapeste, Hungria; e o Museu do Desenho do Tempo em Trancoso, Portugal.

No final dos anos 1980, ele publicou o primeiro livro em todo o mundo sobre arquitetura virtual.

Em 1980 ele deu início à elaboração de *Woiksed* – o primeiro planeta virtual do mundo, antecipando o *Second Life* em cerca de vinte anos. Em 1994, ele recebeu um prêmio europeu por esse projeto.

Nos anos 1990, ele foi curador das primeiras exposições em todo o mundo sobre arquitetura virtual. Uma dessas exposições aconteceu na Bienal de São Paulo 1999 e 2000.

A partir do ano 2000, Emanuel Pimenta dá início aos seus trabalhos e pesquisas em arquitetura espacial, entre os seus projetos nessa área se destacam Kairos, um edifício orbital, e UIRA a primeira Vila Olímpica orbital da história.

Como compositor, desde o final dos anos 1970, tem desenvolvido notações musicais gráficas dentro de ambientes virtuais.

Os seus concertos integrando artes plásticas têm sido realizados em vários países nos últimos trinta anos. Um dos mais importantes momentos foi o seu famoso concerto na Bienal de São Paulo, para quatro grandes formações de músicos, em 1985, lado a lado com John Cage, Robert Rauschenberg, Nan June Paik e Bill Viola entre outros. Em 2008, cria a primeira ópera sobre a Divina Comédia de Dante Alighieri da história da música, com *première* mundial no Festival Abstrata, em Roma, Itália. Em 2009, seu concerto – criado em parceria com o diretor de cinema italiano Dino Viani, que foi responsável pelo filme – tem a sua *première* mundial no Festival Internacional de Cinema de Cannes, França.

Músicos como John Cage, David Tudor, Takehisa Kosugi, John Tilbury, Christian Wolff, Martha Mooke, John DS Adams, Maurizio Barbetti, Michael Pugliese, Umberto Petrin, Susie Georgetis, Audrey Riley e o Manhattan Quartet entre outros têm executado as suas composições.

Colaborou com John Cage, como compositor comissionado, para Merce Cunningham, de 1985 até ao seu desaparecimento em 1992. Continuou como compositor para Merce Cunningham na cidade de Nova Iorque até a sua morte em 2009.

Os seus concertos têm sido executados em alguns dos mais prestigiosos teatros de todo o mundo, como o *Lincoln Center* e *The Kitchen* em Nova Iorque; a *Opera Garnier* ou o *Theatre de La Ville* em Paris; o *Shinjuku Bunka Center* em Tóquio, o teatro Municipal de Montpellier, o Festival de Aix en Provence, o MASP Museu de Arte Moderna de São Paulo, o *La Fenice* em Veneza, e as Bienais de São Paulo e de Veneza entre outros.

Desde 1989, ele tem colaborado regularmente com Lucrezia De Domizio, Baronesa Durini, com incontáveis projetos de arte contemporânea, concertos, exposições e publicações – especialmente sobre Joseph Beuys – em diversos países.

Em 1987, ele começou uma colaboração regular com o legendário filósofo de arte suíço René Berger em numerosos projetos, especialmente na Suíça – com Nan June Paik, Bill Viola, Edgar Morin, Pierre Levy e Basarab Nicolescu entre outros – até a sua morte em 2008.

Na década de 1990, Pimenta desenvolve um gigantesco projeto para uma rede planetária de televisão dedicada à cultura e à arte.

Com mais de quatrocentas composições musicais já gravadas, mais de quarenta compact discs, quatro cd-roms, ele já escreveu e publicou mais de setenta livros, grande parte deles individualmente, vários artigos e uma grande quantidade de livros eletrônicos. Os seus trabalhos têm sido regularmente publicados na Inglaterra, Estados Unidos, Japão, Holanda, Portugal, Brasil, Alemanha, Canadá, Suíça, Hungria, Itália e Espanha.

Ele também tem sido curador para várias instituições, como a Bienal de São Paulo; a Fundação Calouste Gulbenkian e o Centro Cultural Belém, em Portugal, entre outras.

Recebeu o Prêmio Destaque de Markting em 1977, pela Associação Brasileira de Marketing; a Prêmio APCA em 1986, pela Associação Paulista de Críticos de Arte; e o Prêmio Lago Maggiore em 1994 pela UNESCO, AICA Associação Internacional de Críticos de Arte, Conselho de Europa, e Governo Regional da Lombardia. Em 1993 os seus trabalhos são selecionados pela UNESCO, em Paris, como um dos pesquisadores intermédia mais representativos do mundo.

É membro da *SACD – Societè des Autheurs et Compositeurs Dramatiques* em Paris desde 1991.

Ele também é um membro ativo do Tribunal Europeu do Ambiente, em Londres, onde ele tem sido membro da direção desde 1995.

É membro ativo da Academia de Ciências de Nova Iorque; da Sociedade Americana para o Progresso da Ciência, em Washington; da ASMP Sociedade Americana de Fotógrafos de Media; do CAU Conselho de Arquitetura e Urbanismo, no Brasil; da Ordem dos Arquitetos de Portugal. É membro do Comitê Técnico de Arquitetura Espacial do Instituto Americano de Astronáutica e Aeronáutica. É membro consultor da AIVAC - *Association Internationale pour la Video dans les Arts et la Culture*, em Locarno, Suíça. Membro fundador da *International Society for the Interdisciplinary Study of Symmetry – ISIS Symmetry* e da ISA *International Symmetry Association*, ambas em Budapeste.

Foi diretor editorial da revista de arte e cultura RISK ARTE OGGI, em Milão, fundada e dirigida por Lucrezia De Domizio, Baronesa Durini, de 1995 a 2005. Foi também membro da Direção Consultiva Editorial da revista de arte e ciência Forma, em Tóquio, Japão; e é membro do Conselho Editorial da revista de arte e filosofia TechnoEtic Arts, em Bristol, Inglaterra, fundada e dirigida por Roy Ascott.

Emanuel Pimenta tem sido frequentemente convidado, como professor e conferencista, por diversas instituições, entre elas as universidades de Nova Iorque, Georgetown (Washington), Lisboa,

Florença, Lausanne, Tsukuba, São Paulo, Palermo, a Fundação Calouste Gulbenkian em Lisboa, a Fundação Monte Vertia na Suíça, e o Instituto de Tecnologia Technion, em Haifa, Israel.

Ele é fundador e diretor da Academia Holotopia: uma instituição informal orientada à música, arte, filosofia e ciência, na Costa Amalfitana, Itália. É ainda fundador e diretor da Fundação para as Artes, Ciências e Tecnologias – Observatório, em Trancoso, Portugal. Em 2015, foi diretor do Fórum Mundial da Arte, em Bolognano, Itália.

Ele vive entre Locarno, Suíça, que é a sua residência principal, Nova Iorque e Lisboa. O seu site na Internet é o www.emanuelpimenta.net. Nesse site é possível encontrar seus projetos de arquitetura, projetos musicais, livros, livros eletrônicos, projetos fotográficos e ensaios, artigos e filmes.

alguns livros de Emanuel Pimenta

Sr. Chico - Um Mestre Zen nas Florestas

John Cage - O Silêncio da Música

Koellreutter - As Revoluções Musicais de um Mestre Zen

Medauar - O Homem que Sabia Demais

John Cage - Koan da Nºao Violência

John Cage: Como Mudar o Mundo

One Hundred Years with John Cage

Virtual Notations (Music)

KAIROS - Um pássaro em Órbita da Terra (arquitetura espacial)

30 Years of Architecture

UIRA Vila Olímpica Orbital (arquitetura espacial)

Logical Traps

Arte e Zen

Seres Ocultos

Sociedade Low Power

Mondo: Literatura e Democracia

Walden Zero Project

Kirkos: A Dialogue Between Marcel Duchamp and Josqin Des Près

Neapolis

Firenze: Mind Battle Fields of a Magical City

Ascona